Strategic Learning Package for

Physiology of Behavior

Strategic Learning Package

for

Carlson

Physiology of Behavior

Sixth Edition

prepared by

Mary Carlson

Kerstin Carlson Le Floch

Neil R. Carlson
University of Massachusetts at Amherst

Allyn and Bacon
Boston · London · Toronto · Sydney · Tokyo · Singapore

ISBN 0-205-27435-8

Printed in the United States of America

10 9 8 7 6 5 4 3 2   02 01 00 99 98

Table of Contents

Preface

Welcome to your textbook and to your study guide. We remember what it was like to be students, and we have tried to write something that will help you learn the information in the text in a way that it will stay with you. Your goal is to learn something interesting and useful, and our goal is to help you accomplish that.

The purpose of a study guide is, as its name suggests, to guide you through the text and make sure you see the important points, think about them, write about them, and, most important, *remember* them. We hope that you will remember what you learn even after the semester is over.

Learning is an active process, not a passive one. It is easy to read a chapter and say to yourself, "Yes, I understand that," but still learn and remember very little. If you do not put some effort into your studying, you are not spending your time very effectively.

Although you already have some well established study habits, we urge you to think about trying something different this semester. Begin by leafing through the chapter so that you can see what it is about. Look at the outline at the beginning of the chapter, then read the chapter. Don't take notes or worry about remembering too many details—you only want to see what the chapter is all about. Do all of this *before your professor lectures about the material.* If you are familiar with the chapter, you will understand the lecture much better, and you will remember much more of what you hear. Your lecture notes will also be much clearer.

Now you are ready to work with the study guide. With the exception of Chapter 1 (which is short) we have divided each chapter into two lessons. Finish the first lesson and take a break. You will find it easier to work if you have a break to look forward to. Take some time off, even if you feel that you could finish the chapter in one sitting. As the semester goes on, you will find it easier to begin studying if the habit of taking a break is well established.

Each lesson in the study guide is broken into learning objectives, which are enclosed in a box. The objective states what you should learn to do. Before the first objective, you will see the following statement: "Read the interim summary on page 000 of your text to re-acquaint yourself with the material in this section." Do just that. The interim summaries will remind you about what you read previously. Then, after the objective, you will find the following statement: "Read

pages 000-000 and answer the following questions." Do that, too. Then, begin answering the questions. *Try to answer each question without looking at the book.* If all you do is to copy the answer from the book into the study guide, you will learn very little. If the purpose of having a study guide were simply to have a list of answers to a list of questions, we could have written the answers, too. Don't be a scribe, copying from the book to the study guide. That's boring and pointless. If you can't answer a particular question, read the appropriate section in the text again, close the book (or at least push it away so that you can't see it), and then write the answer from memory. *If you cannot remember an answer a short time after reading the information in the book, how can you expect to remember it when you take the exam?* The delay between reading the information and writing it in your own words is crucial to your remembering what you have read.

The study guide contains some features that will help you to study your text. It lists the pages that you should read, and many questions refer to specific figures that contain information that you will need to answer a question. We have also included a set of concept cards. These cards contain most of the terms that are listed in the margins of each chapter. Learning these terms will help you acquire the vocabulary of physiological psychology. To use the concept cards effectively, cut them apart, put them in a stack, and test yourself. Obviously, you do not need to learn the definitions word-for-word, because there is more than one way to explain a term. But if you forget something important about a term, put the card into a stack that you will go through again until you are satisfied with your performance.

We have tried to supply you with the tools to change some neural circuits in your brain—after all, that is what learning is all about. Let us know how the study guide works for you. If you have some suggestions for the next edition, please write to us at the address listed in the preface of the textbook.

Good luck!

Mary Carlson
Kerstin Carlson Le Floch
Neil Carlson

How to Use the Computerized Study Guide

The diskettes included with the Strategic Learning Package contain computerized exercises to help you learn the material in the text.

Requirements:

PC-compatible computer running Windows 3.1 or Windows 95, hard disk with 4.5 megabytes available, VGA (or better) display with 256 colors. *Macintosh version also available: See note at the end of this section.*

Windows 3.1 Installation:

1. Have Windows 3.1 running.

2. Insert diskette 1 in drive A: (or B:).

3. With the mouse, click on **File** at the top of the screen, then click on **Run** on the menu that appears. (Alternatively, you can press the "Alt" and "f" keys together, then release them and press the "r" key.)

4. A dialog box entitled "Run" will appear on your screen, requesting you to type the **Command** line. Type **a:setup** (or **b:setup**) and press <enter>.

5. In a few seconds, your screen will turn blue and you will see the SGUIDE for Windows setup screen.

6. To continue, press <enter> or click on **Continue**.

7. The program will normally be loaded in a directory called SGUIDE, which will be created on your hard disk. If you want the program to go into another directory, type the path you want to use. (For example, if your hard disk is D: and you want to create a directory called STUDY, you would type D:\STUDY.

8. Press <enter> or click on **Continue**.

9. The setup program will load SGUIDE and its associated files onto your hard disk. Follow the instructions when it asks you to remove disk 1 and insert disk 2.

10. If everything goes well, you will see a dialog box saying, "The setup process was successful!" Press <enter> or click on **OK**.

11. A new icon, labeled "Computerized Study Guide," will have been added to your Application program group. If you want to, you can move the icon to another group by clicking on it with the mouse and dragging it elsewhere on the desktop. (Consult your Windows users' manual for further information on arranging icons on the desktop.)

Windows 95 Installation:

1. Have Windows 95 running.

2. Insert diskette 1 in drive A: (or B:).

3. With the mouse, click on **Start,** then click on **Run**. The dialog box will ask for the name of the program you want to open. Type **a:setup** (or **b:setup**).

4. In a few seconds, your screen will turn blue and you will see the SGUIDE for Windows setup screen.

5. To continue, press <enter> or click on **Continue**.

6. The program will normally be loaded in a directory called SGUIDE, which will be created on your hard disk. If you want the program to go into another directory, type the path you want to use. (For example, if your hard disk is D: and you want to create a directory called STUDY, you would type D:\STUDY.

7. Press <enter> or click on **Continue**.

8. The setup program will load SGUIDE and its associated files onto your hard disk. Follow the instructions when it asks you to remove disk 1 and insert disk 2.

9. If everything goes well, you will see a dialog box saying, "The setup process was successful!" Press <enter> or click on **OK**.

10. A new icon, labeled "Computerized Study Guide," will have been added to a program group called "Applications." If you want to, you can move the icon to the desktop by clicking on it with the mouse and dragging it there. Then you can close the Application group and the Program Manager. (Consult your Windows users' manual for further information on arranging icons on the desktop.)

Running SGUIDE:

1. Double click on the icon.

2. The opening screen announces the name of the program. Click on "Continue."

3. The menu lists all of the chapters in the book. Choose one by clicking on it.

4. A list of exercises will appear at the bottom of the list. Choose one by clicking on it.

5. If you choose **Terms and Definitions**:

   a. You will be presented with a definition and must type the appropriate term. Correct any typographical errors by using the backspace and retyping the correct letters, then press <enter>.

   b. The computer will respond "**Correct**" or "**Incorrect**," and make appropriate noises (if your computer has a sound card). If your response was correct, simply press <enter>, and the next item will be presented. If your response was incorrect, the computer will present the correct response.

   Important note: If you misspell the term, the computer will say that the answer was incorrect—even if the word was long and complicated and you got only one letter wrong. In some cases, there might even be more than one correct answer. A human reading what you type might say, "Yes, that's a valid answer, too"—and mark your answer "correct." But when I wrote this program I couldn't anticipate all the possible correct answers. Please don't get mad at the program, but recognize its limitations.

   c. When you finish the questions, the computer will start another round and present the items you answered incorrectly. If you get all the answers right, the computer will let you go back to the menu.

6. If you choose Self Tests:

   a. You will be presented with multiple choice questions. The first part of the question appears in a box at the top of the screen, and the four choices appear below. Click on the correct alternative. (You can click anywhere along the line.)

   b. The computer will respond "**Correct**" or "**Incorrect**," and make the appropriate sound. If your response was correct, simply press <enter> and the next item will be presented. If your response was incorrect, choose another response.

   c. When you finish the questions, the computer will start another round and present the items you answered incorrectly. If you get all the answers

right, the computer will let you go back to the menu.

7. If you choose Figures and Diagrams:

   a. You will be presented with a figure, which you will probably recognize from your text. You will also see a set of labels enclosed in boxes scattered to the right and/or top of the screen. Your job is to click on each of the labels with the mouse and drag them to their appropriate locations.

   b. The first time you try this exercise, click on **Help**, and you will see an exercise that explains how to move the labels.

   d. When you move a box to the correct location, the computer will make a noise and the background of the box will change its color. When all labels are correctly placed, you can go on to the next figure. Even if you have not finished a figure, you can click on **Next Figure**, which does just what you'd expect.

8. For *all* choices:

   At any time, you can click on **EXIT** and return to the menu—or, if you are already at the menu, you can quit the program.

9. I hope you enjoy the program and—more importantly—that it helps you learn. If you have any questions, criticisms, or suggestions for future versions, please let me know. My e-mail address is nrc@psych.umass.edu and my postal addresses is Department of Psychology, Tobin Hall, University of Massachusetts, Amherst, MA 01003, USA.

Macintosh Users

No, I haven't forgotten you. I wrote a Macintosh version of this program. In fact, I bought a Mac so I could do so. To receive the Mac version, please send your Windows disks to:

Joyce Nilsen
Allyn and Bacon
160 Gould Street
Needham Heights, MA 02194

You will receive a new set of disks by return mail.

Neil Carlson

CHAPTER 3
Structure of The Nervous System

Lesson I: Basic Features of the Nervous System and Some Structures of the Central Nervous System

Read the interim summary on pages 64-65 of your text to re-acquaint yourself with the material in this section.

Learning Objective 3-1   Describe the appearance of the brain and the terms used to indicate directions and planes of section.

Read pages 57-59 and answer the following questions.

1. Draw two pictures of a snake. Make the first a side view and the second a front view. Label your drawings with these terms of anatomical direction: neuraxis, anterior, posterior, rostral, caudal, dorsal, ventral, lateral, and medial. (Study Figure 3.1 in your text.)

2. Now repeat this exercise by labeling two stick figure drawings of a human.

3. Define these frequently used terms in your own words.

    a. ipsilateral

    b. contralateral

4. To confirm your understanding of the nomenclature for planes of section, try using these terms to describe these pieces of food: *cross* or *frontal, horizontal,* and *sagittal.* You will have to imagine the neuraxis. (Study Figure 3.2 in your text.)

_____ cutting a hamburger bun into top and bottom halves

_____ a slice of bread

_____ slicing a fish into two symmetrical halves

Learning Objective 3-2   Describe the blood supply to the brain, the meninges, the ventricular system, and flow of cerebrospinal fluid through the brain and its production.

Read pages 59-64 and answer the following questions.

1. Name the two major divisions of the nervous system and then list their parts. (See Table 3.1 and Figure 3.3 in your text.)

    1.

    2.

2. a. Approximately how much of the blood flow from the heart does the brain receive?

    b. Why is an uninterrupted blood supply essential?

    c. What are the consequences if blood supply is interrupted for 1-second? 6-seconds? a few minutes?

3. a. Name and describe the blood vessels through which blood flows after it leaves the heart.

    b. List the major sets of arteries in the brain and identify the areas they serve. (See Figure 3.4 in your text.)

        1.   2.

    c. How is the normal flow of blood affected if a blood vessel becomes blocked?

4. a. Return to Figure 3.3 in your text and then list and describe the three layers of the meninges beginning with the outer layer.
        1.
        2.
        3.

    b. What area lies between the arachnoid membrane and the pia mater?

    c. What liquid fills this space?

d. Which two layers of the meninges fuse outside the CNS? What does this sheath cover?

5. Describe how the CSF protects the brain.

6. Define *ventricles* in your own words.

7. Study Figure 3.5 in your text and describe the major components of the ventricular system.

   a. Name the two largest chambers in the brain.

   b. What structure

      1. crosses the middle of the third ventricle?

      2. connects the third and fourth ventricles?

8. Where is CSF produced and how is it reabsorbed?

9. Write a sentence describing the flow of CSF, using all of the following terms.

   lateral ventricles, third ventricle, fourth ventricle, blood supply, choroid plexus, arachnoid granulations, small openings, subarachnoid space, cerebral aqueduct, superior sagittal sinus

10. What is the cause and treatment of obstructive hydrocephalus? (See Figure 3.6 in your text.)

11. To review: The interim summary presents important anatomical nomenclature, the divisions of the central and peripheral nervous system, and an overview of the ventricular system and cerebrospinal fluid.

Read the interim summary on page 79 of your text to re-acquaint yourself with the material in this section.

Learning Objective 3-3   Outline the development of the central nervous system.

Read pages 65-67 and answer the following questions.

1. a. During development of the central nervous system, where are the cells that give rise to neurons located? (Study Figure 3.7 in your text.)

   b. In which direction do newly formed neurons migrate?

   c. By what means do they reach their final location?

2. Study Figure 3.8 in your text and then label the portions of the embryonic hollow tube shown in Figure 1a that become the forebrain, the midbrain, and the hindbrain.

3. On Figure 1b, indicate how the forebrain continues to develop by labeling the telencephalon, the diencephalon, the mesencephalon, and the two parts of the hindbrain—the metencephalon and the myelencephalon.

4. On Figure 1c, indicate the location of the cerebral hemisphere, cerebellum, thalamus, hypothalamus, pituitary gland, midbrain, pons, , medulla, brain stem and spinal cord.

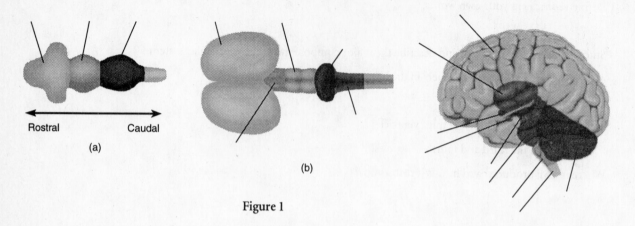

Rostral   Caudal

(a)

(b)

Figure 1

(c)

5. Immediately review what you have just learned by studying Table 3.2 in your text. Now complete the blank version of that table, below.

Anatomical Subdivisions of the Brain			
Major Division	*Ventricle*	*Subdivision*	*Principal Structures*

6. a. Describe how neurons continue to develop and form synaptic connections with postsynaptic cells after they reach their final locations. Be sure to mention the role of the postsynaptic cell.

   b. What happens to about half of the neurons produced in the neural tube, which fail to form synaptic connections?

c. Outline a possible explanation for this phenomenon.

Learning Objective 3-4   Describe the telencephalon, one of the two of the two major structures of the forebrain.

Read pages 67-73 and answer the following questions.

1.   Name the two major components of the forebrain.

    1.   2.

2.   The telencephalon includes most of the two _____ _____, which are covered by the

    _____ _____.

3.   List two subcortical structures of the telencephalon that are found deep within the cerebral hemispheres.

    1.   2.

4.   a. Large grooves in the cerebral cortex are called _____ and small ones are called

        _____. The bulges between grooves are called _____. (See Figure 3.9 in your text.)

    b. Why is the human cerebral cortex convoluted rather than smooth?

    c. What do the following terms apply to, and what is responsible for the colors?

        1. gray matter

        2. white matter

5.   Study Figure 3.10 in your text and describe the location of these three areas of cerebral cortex noting nearby fissures or sulci.

    1. primary visual cortex

    2. primary auditory cortex

    3. primary somatosensory cortex

6.   What kind of sensory information is not sent to the contralateral side of the brain?

7.   In general, what kinds of functions are mediated by the anterior and posterior regions of the cerebral cortex?

8.   Label the following regions on Figure 2 on the next page: the four lobes; the primary visual, somatosensory, auditory, and motor cortex; and the central sulcus. Indicate rostral, caudal, dorsal, and ventral, making a total of 13 items to label. Study Figures 3.10 and 3.11 in your text.

9.   Compare the functions of sensory association cortex located close to and farther away from primary cortex.

10. What are the general functions of the motor association cortex? the prefrontal cortex?

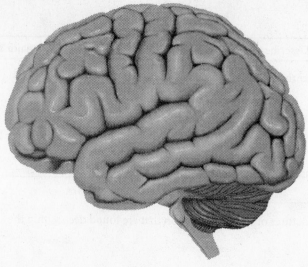

Figure 2

11. The functions of the cerebral hemispheres are _____. The right hemisphere is involved in the

_____ of information and the left is involved in the _____ of information.

12. a. Study Figure 3.11 and 3.12 in your text. In which region is the neocortex found? the limbic cortex: the cingulate gyrus?

b. Name and describe the function of the largest commissure in the brain.

13. Label the regions indicated by the lines on Figure 3.

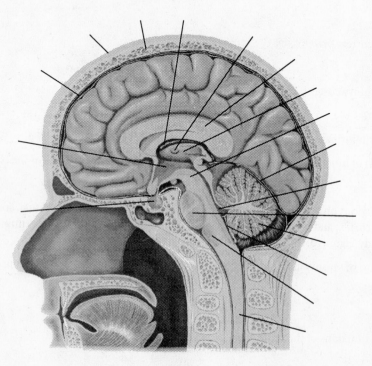

Figure 3

14. Study Figure 3.13 in your text and list the three most important structures that are a part of the limbic system.

    1.   3.

    2.

15. What kinds of functions are controlled by the limbic system?

16. What structure connects the hippocampus with other parts of the brain?

17. a. List the major parts of the basal ganglia. (Figure 3.14 in your text shows their location.)

    1.   3.

    2.

    b. What is the general function of the basal ganglia?

    c. Note the cause and some of the symptoms of Parkinson's disease.

Lesson I Self Test

1. As you look down on the snake, you see its _____ surface, but it slithers along the ground on its _____surface.

    a. lateral; ventral
    b. ventral; medial
    c. dorsal; ventral
    d. dorsal; lateral

2. Which of the following would not be visible in a midsagittal view of the brain?

    a. the lateral fissure
    b. the limbic cortex
    c. the corpus callosum
    d. the cingulate gyrus

3. The meninges and subarachnoid space surround the brain in the following order, beginning with the outer layer:

    a. dura mater, pia mater, arachnoid membrane, subarachnoid space
    b. dura mater, arachnoid membrane, subarachnoid space, pia mater
    c. pia mater, dura mater, subarachnoid space, arachnoid membrane
    d. pia mater, arachnoid membrane, subarachnoid space, dura mater

4. Cerebrospinal fluid is produced by the

    a. meninges.
    b. subarachnoid space.
    c. choroid plexus.
    d. ventricles.

5. After cerebrospinal fluid has circulated through the brain and subarachnoid space it is

    a. excreted by the kidneys.
    b. recirculated for the next three hours.
    c. reabsorbed into the blood supply.
    d. transported through the central canal to the abdomen.

6. In the central nervous system neurons develop

    a. on the inner surface of the embryonic hollow tube.
    b. in radially oriented glial cells.
    c. in the chambers which become the ventricles of the brain.
    d. in cortex of the brain.

7. The three major parts of the brain are the

    a. telencephalon, the diencephalon, and the metencephalon.
    b. cerebral cortex, the association cortex, and the brain stem.
    c. frontal lobe, the parietal lobe, and the temporal lobe.
    d. forebrain, the midbrain, and the hindbrain.

8. Neurons that do not establish synaptic connections with a postsynaptic cell

   a. migrate to the bone marrow.
   b. may have an insufficient number of growth cones.
   c. may later become malignant.
   d. eventually die.

9. The lateral fissure separates the _____ lobe from the _____ lobe.

   a. frontal; parietal
   b. frontal; temporal
   c. temporal occipital
   d. parietal; occipital

10. The corpus callosum connects the

    a. two hemispheres of the brain.
    b. structures of the limbic system.
    c. pituitary gland and the hypothalamus.
    d. thalamus and the hypothalamus.

11. The most important structures of the limbic system are the limbic cortex, the

    a. hippocampus, and the amygdala.
    b. basal ganglia, and the thalamus.
    c. primary motor cortex, and the primary association cortex.
    d. hypothalamus, and the pituitary gland.

12. The limbic system plays a role in

    a. planning and execution of movement.
    b. visual and auditory functions.
    c. control of the endocrine system.
    d. emotional behavior, learning, and memory.

Lesson II: Some Structures of the Central Nervous System and the Peripheral Nervous System

Read the interim summary on page 79 of your text to re-acquaint yourself with the material in this section.

Learning Objective 3-5   Describe the two major structures of the diencephalon.

Read pages 73-76 and answer the following questions.

1. a. The two most important structures of the diencephalon are the _____ and the _____.

   b. Study Figures 3.14 and 3.17 in your text and describe the location and appearance of the thalamus. Be sure to mention the massa intermedia in your answer.

2. Define *projection fibers* and *nuclei* in your own words and explain their relation to the function of the thalamus.

3. Where do these thalamic nuclei receive and relay their sensory information?

   a. lateral geniculate nucleus

   b. medial geniculate nucleus

   c. ventrolateral nucleus

4. a. Describe the location and general functions of the hypothalamus. (Study Figures 3.14 and 3.15 in your text.)

   b. What structure is attached to the base of the hypothalamus? What other structure lies just in front of it? (See Figures 3.15 and 3.16 in your text.)

c. The hypothalamus controls the anterior pituitary gland, and through it most of the _____

   system, by secreting the hypothalamic _____, which are produced by specialized neurons called

   _____ _____, located at the base of the _____ _____.

d. To explain why the anterior pituitary gland is often called the "master gland," briefly summarize the effects of some of the hormones it secretes.

e. Describe the production and secretion of the hormones of the posterior pituitary gland.

f. Now summarize the effects of some of the hormones secreted by the posterior pituitary gland.

Learning Objective 3-6   Describe the two major structures of the midbrain, the two major structures of the hindbrain, and the spinal cord.

Read pages 76-79 and answer the following questions.

1. List the two major parts of the midbrain or mesencephalon.

   1.   2.

2. Then list the two principal structures of the tectum. (See Figure 3.17 in your text.)

   1.   2.

3. Continue by listing the structures of the brain stem.

   1.   2.   3.

4. Finally, list the principal structures of the midbrain shown in Figure 3.17d and noted in the marginal definitions on page 76 in your text.

   1.   4.

   2.   5.

   3.   6.

5. a. Briefly describe the composition and appearance of the reticular formation.

   b. List some of its functions.

6. a. What gives the periaqueductal gray matter its name?

   b. List a function of the

      1. periaqueductal gray matter.

      2. red nucleus.

      3. substantia nigra.

7. a. List the two divisions of the hindbrain.

      1.   2.

b. Now list the two structures of the metencephalon.

1.   2.

8. Briefly describe the location of the cerebellar

a. cortex.

b. deep nuclei.

c. peduncles. (See Figure 3.17.)

9. Briefly describe some of the functions of the cerebellum and the kinds of disability resulting from damage to this structure.

10. Describe the location of the pons and the function of several nuclei found there. (Look back at Figures 3.12 and 3.17 in your text.

11. Describe the location and functions of the medulla oblongata, the major structure of the myelencephalon.

12. What is the principal function of the spinal cord?

13. How is the spinal cord protected?

14. Name the passage at the center of each vertebra through which the spinal cord passes. (See Figure 3.18 in your text.)

15. a. Explain why the spinal cord is only about two thirds as long as the vertebral column.

b. Describe the cauda equina and explain how it is possible to eliminate sensations from the lower part of the body using a caudal block. (Refer to Figure 3.3 in your text.)

c. Study Figure 3.19 in your text and describe the dorsal roots and ventral roots and the axons they contain. Be sure to notice the location of the white matter and gray matter.

Read the interim summary on page 85 of your text to re-acquaint yourself with the material in this section.

Learning Objective 3-7   Describe the peripheral nervous system, including the two divisions of the autonomic nervous system.

Read pages 80-85 and answer the following questions.

1. List the two sets of nerves of the peripheral nervous system and describe their general functions.

1.   2.

2. Look back at Figure 3.3 in your text and outline the spinal nerve pathways throughout the body.

3. The cell bodies of _____ axons which bring sensory information into the brain and spinal cord are located _____ the central nervous system with the exception of the _____ _____ which is a part of the brain. Incoming axons are called _____ axons and the cell bodies from which they arise are found in the _____ _____ _____.
    _____ axons leave the spinal cord through the _____ _____ and control the muscles and glands. (See Figure 3.20 in your text.)

4. Study Figure 3.21 in your text and list the names, numbers, and functions of at least four of the twelve pairs of cranial nerves that leave the brain. Begin with the vagus nerve.

    1. vagus nerve (10)

    2.

    3.

    4.

5. List the two main divisions of the PNS and identify their functions.

    1.   2.

6. The autonomic nervous system is further divided into two parts. List them.

    1.   2.

7. a. Using examples, describe the kinds of activity mediated by the sympathetic division.

    b. Where are the cells bodies of sympathetic motor neurons located and where do their axons exit the CNS?

    c. Where do these axons go?

    d. Study Figure 3.22 in your text and describe how the sympathetic ganglion chain is formed.

8. a. Axons of the ANS that leave the spinal cord are called _____ _____ and, with one exception, enter the ganglia of the _____ _____. The exception is the _____ _____ located in the center of the adrenal gland.

    b. Where do postganglionic neurons send their axons?

c. Which hormones are secreted by the cells of the adrenal medulla and what are their effects on the body?

9. Using examples, describe the kinds of activity mediated by the parasympathetic division.

10. Name the two regions that give rise to preganglionic axons of the parasympathetic division.

    1.   2.

11. To review: Study Table 3.3 in your text.

Lesson II Self Test

1. The _____ surrounds the third ventricle and its two most important structures are the _____.
    a. forebrain; the telencephalon and the diencephalon
    b. diencephalon; thalamus and the hypothalamus
    c. limbic system; basal ganglia and the amygdala
    d. diencephalon; fornix and the massa intermedia

2. The thalamus is responsible for
    a. most of the neural input received by the cerebral cortex.
    b. emotional behavior.
    c. movement of a particular part of the body.
    d. behaviors related to survival of the species.

3. Neurons in the hypothalamus
    a. control the peripheral nervous system.
    b. send projection fibers through the optic chiasm.
    c. are controlled by hormones secreted by the anterior pituitary gland.
    d. are involved behaviors such as fighting and fleeing.

4. The anterior pituitary gland produces
    a. vasopressin.
    b. oxytocin.
    c. gonadotropic hormones.
    d. estrogen.

5. The principal structures of the tectum are the
    a. superior and inferior colliculi.
    b. hippocampus and amygdala.
    c. thalamus and hypothalamus.
    d. lateral and medial geniculate nuclei.

6. The reticular formation
    a. relays visual information from the retina to the rest of the brain.
    b. appears as four bumps on the brain stem.
    c. is one of two major fiber systems within the brain.
    d. plays a role in sleep and arousal.

7. The periaqueductal gray matter is so called because of an abundance of
    a. fibers.
    b. cell bodies.
    c. synapses.
    d. axons.

8. The spinal cord is _____ the vertebral column.
    a. fused to
    b. longer than
    c. outside
    d. shorter than

9. Dorsal roots contain _____ axons and ventral roots contain _____ axons.
    a. unipolar; bipolar
    b. myelinated; unmyelinated
    c. afferent; efferent
    d. motor; sensory

10. The cranial nerve that regulates the function of organs in the thoracic and abdominal cavities is the
    a. preganglionic nerve.
    b. hypoglossal nerve.
    c. vagus nerve.
    d. trigeminal nerve.

11. The two divisions of the autonomic nervous system are the

a. brain and the spinal cord.
b. somatic nervous system and the autonomic nervous system.
c. sympathetic division and the parasympathetic division.
d. spinal nerves and the cranial nerves.

12. _____ _____ leave the spinal cord through the ventral root.

a. Preganglionic axons
b. Postganglionic axons
c. Sympathetic ganglia
d. Cranial nerves

Answers for Self Tests

Lesson I

1. c Obj. 3-1
2. a Obj. 3-1
3. b Obj. 3-2
4. c Obj. 3-2
5. c Obj. 3-2
6. a Obj. 3-3
7. d Obj. 3-3
8. d Obj. 3-3
9. b Obj. 3-4
10. a Obj. 3-4
11. a Obj. 3-4
12. d Obj. 3-4

Lesson II

1. b Obj. 3-5
2. a Obj. 3-5
3. d Obj. 3-5
4. c Obj. 3-5
5. a Obj. 3-6
6. c Obj. 3-6
7. b Obj. 3-6
8. d Obj. 3-6
9. c Obj. 3-6
10. c Obj. 3-7
11. c Obj. 3-7
12. a Obj. 3-7

CHAPTER 4
Psychopharmacology

Lesson I: Principles of Psychopharmacology and Sites of Drug Action

Read the interim summary on page 94 of your text to re-acquaint yourself with the material in this section.

Learning Objective 4-1 Describe the routes of administration and the distribution of drugs within the body.

Read pages 87-91 and answer the following questions.

1. What is a generally accepted definition of a drug?

2. Define the terms *drug effects* and *sites of action*.

3. The process by which drugs are absorbed, distributed within the body, metabolized, and excreted is referred to as

   _____.

4. For laboratory animals, the most common route of drug administration is injection. List and briefly describe the four routes of injection.

   1.

   2.

   3.

   4.

5. What is the most common route used to administer drugs to humans? Why can some drugs not be administered this way?

6. List and describe four other routes used to administer drugs to humans.

   1.

2.

3.

4.

7. Why does the lipid solubility of a drug affect the rate at which the drug reaches the sites of action in the brain?

8. Name and describe two ways in which drugs can be administered directly into the brain.

9. Describe the rapidity with which cocaine enters the blood after intravenous injection, oral administration, smoking, and sniffing. (See Figure 4.1 in your text.)

10. a. Describe the phenomenon known as depot binding.

    b. What source of depot binding is found in the blood? How does it interact with the absorption of drugs? (See Figure 4.2 in your text.)

    c. What are some other sources of depot binding?

11. Drugs do not remain in the body indefinitely. What eventually happens to them?

| *Learning Objective 4-2*   Describe drug effectiveness, the effects of repeated administration, and the placebo effect. |

Read pages 91-94 and answer the following questions.

1. Study Figures 4.3 and 4.4 in your text and explain the following terms:

    a. dose-response curve

    b. margin of safety

    c. therapeutic index

2. Provide two reasons why drugs vary in their effectiveness.

3. Name two phenomena that may occur after repeated administration of a drug.

4. What are withdrawal effects? Why do they occur?

5. Describe two types of compensatory mechanisms.

6. Using barbiturates as an example, describe some of the dangers of drug tolerance.

7. Review research by Levine et al. (1979) on the placebo effect.

    a. What was the effect of a placebo pain killer on subjects' sensitivity to pain?

    b. What was the effect of an injection of naloxone on pain sensitivity?

    c. Hence, what can we conclude about the effects of subjects' belief that they had received a pain killer? (See Figure 4.5 in your text)

Read the interim summary on pages 98-99 of your text to re-acquaint yourself with the material in this section.

> *Learning Objective 4-3*   Describe sites of drug action and the production, storage, and release of transmitter substances.

Read pages 94-98 and answer the following questions.

1. Most drugs that affect behavior do so by affecting synaptic transmission. Those that block or inhibit the postsynaptic effects are called _____. Those that facilitate them are called _____.

2. Review the sequence of synaptic activity that you learned about in Chapter 2.

Figure 1 on the next page illustrates eleven ways that drugs can affect synaptic transmission. As you answer the following questions, fill in the missing information. (See Figure 4.6 in your text.)

3. Describe the two ways in which drugs can affect the production of transmitter substance. Put your answers in boxes 1 and 2 and indicate if the effects are those of an agonist or antagonist.

4. What effect can drugs have on the storage of transmitter substance in synaptic vesicles? How? Explain the process below and write a four word summary in box 3 and indicate if the drug is an agonist or antagonist.

5. What effect can drugs have on the release of transmitter substance? Explain the process, and write a summary statement in boxes 4 and 5. Indicate if the drug is an agonist or antagonist.

6. Describe how a drug acts as a direct agonist. Write a summary statement in box 6.

7. How do direct antagonists function? What is another name for drugs with this effect? Write a summary statement in box 7.

8. Describe the effects of an inverse agonist. Include the term *noncompetitive binding*. (See Figure 4.7 in your text.)

9. In contrast, how does an indirect agonist function? (See Figure 4.7.)

10. Drugs can stimulate or block presynaptic autoreceptors. Describe these effects and write summary statements in boxes 8 and 9.

11. What two processes terminate the postsynaptic potential? How do drugs affect these processes? Are they agonists or antagonists? Write summary statements in boxes 10 and 11. (Note that the answer to this question is found on page 98 in your text, in the last paragraph of this section.)

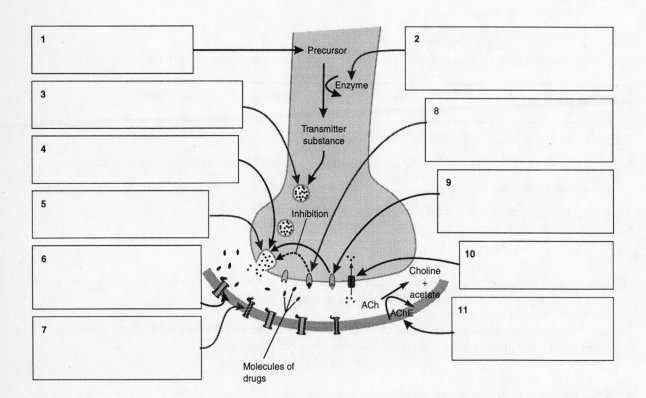

Figure 1

12. Let's review how drugs can affect axoaxonic synapses—synapses of one terminal button with another. (Turn back to page 43 in your text.)

    a. What are presynaptic heteroreceptors?

    b. How do these receptors produce presynaptic inhibition? (See Figure 4.8 in your text.)

    c. How do they produce presynaptic facilitation? (See Figure 4.8.)

    d. What effect can drugs have on presynaptic heteroreceptors?

13. Describe the agonistic and antagonistic effects that drugs can have on dendritic autoreceptors. (See Figure 4.9 in your text.)

14. The table below summarizes the effects of drugs that bind with receptors. Fill in the blanks. (Try not to simply copy the answers from Table 4.1 in your text.)

Site of action	Effect of activated receptor	Effect of drug on receptor	Effecton synaptic transmission (agonist or antagonist)
Postsynaptic receptor			
Presynaptic autoreceptor			
Presynaptic heteroreceptor			
Presynaptic heteroreceptor			
Dendritic autoreceptor			

Lesson I Self Test

1. Which of the following routes of drug administration is not used for laboratory animals?

    a. intravenous injection
    b. subcutaneous injection
    c. sublingual administration
    d. intracerebroventricular administration

2. Which of the following is not true of depot binding?

    a. It can delay the effects of a drug.
    b. It can prolong the effects of a drug.
    c. Albumin is a source of depot binding.
    d. Kidneys are a source of depot binding.

3. The therapeutic index of a drug is

    a. a measure of a drug's margin of safety.
    b. the dose of a drug that produces toxic effects in 50 percent of animals.
    c. very low for drugs such as Valium.
    d. very high for barbiturates.

4. Which of the following is not true about repeated administration of a drug?

    a. It can inhibit tolerance.
    b. It can produce sensitization.
    c. Withdrawal symptoms can appear after a person stops taking the drug.
    d. Compensatory mechanisms may be engaged.

5. The placebo effect

    a. shows that the mind can affect behavior without involving the brain.
    b. is especially important in animal research.
    c. has been analyzed through the use of drugs.
    d. only occurs in people who can be hypnotized.

6. Select the incorrect statement.

    a. When a drug acts as a precursor, it increases the production of a neurotransmitter, serving as an agonist.
    b. If a drug inactivates enzymes responsible for the production of a neurotransmitter, it acts as an agonist.
    c. Transporter molecules that fill synaptic vesicles may be blocked by a drug, which then serves as an antagonist.
    d. Some antagonist drugs prevent the release of transmitter substance from terminal buttons.

7. A drug that mimics the effects of a transmitter substance acts as a

    a. receptor blocker.
    b. direct antagonist.
    c. direct agonist.
    d. indirect ligand.

8. Drugs that bind with postsynaptic receptors can serve as antagonists when they

    a. cause ion channels to open.
    b. depolarize cations.
    c. prevent the synaptic vesicles from releasing transmitter substance.
    d. bind with the receptors but do not open the ion channels.

9. Select the incorrect statement

    a. Receptor blockers or direct antagonists refer to the same process.
    b. Noncompetitive binding means that a molecule does not compete with molecules of the transmitter substance for the same binding site.
    c. A drug that binds noncompetitively can act as an inverse antagonist.
    d. Drugs that block presynaptic autoreceptors increase the release of the transmitter substance.

10. Which of the following is not true of presynaptic heteroreceptors?

    a. They are contained in the first (presynaptic) terminal button of axoaxonic synapses.
    b. Drugs can make them insensitive to the transmitter substance.
    c. They can inhibit the opening of voltage-dependent calcium channels.
    d. They can facilitate the opening of voltage-dependent calcium channels.

11. Drugs that bind with and block dendritic autoreceptors

    a. produce an inhibitory hyperpolarization.
    b. inhibit production of the transmitter substance.
    c. disrupt the regulatory mechanism that controls the production of transmitter substance.
    d. act as agonists.

12. Termination of the postsynaptic potential

    a. can occur when molecules of the transmitter substance are taken back into the dendrite through reuptake.

    b. can occur when molecules of the transmitter substance are destroyed by an enzyme.

    c. may involve the enzyme choline acetyltransferase.

    d. may be inhibited by drugs that act as antagonists.

Lesson II: Neurotransmitters and Neuromodulators

Read the interim summary on page 118 of your text to re-acquaint yourself with the material in this section.

> *Learning Objective 4-4*   Review the general role of neurotransmitters and neuromodulators, and describe the acetylcholinergic pathways in the brain and the drugs that affect these neurons.

Read pages 99-102 and answer the following questions.

1. In the brain, most synaptic activity is accomplished by two transmitter substances, one with _____ effects (_____) and one with _____ effects (_____).

2. List three sites in the body where synapses release acetylcholine. (See Figure 4.10 in your text.)

    1.   3.

    2.

3. Why was acetylcholine the first neurotransmitter to be discovered?

4. Describe some of the functions influenced by acetylcholinergic neurons in the brain.

5. a. Figure 2 illustrates the chemical reactions responsible for the production of acetylcholine. Label each component. (Study Figure 4.11 in your text.)

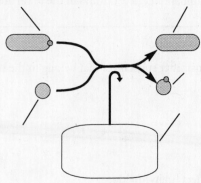

Figure 2

    b. Explain the details of this reaction in your own words.

6.   List two drugs that affect the release of acetylcholine and describe their effects.

1.

2.

7.   a. Describe the symptoms of a hereditary disorder called myasthenia gravis.

b. Identify the AChE inhibitor that is used to treat this disorder and explain its effects.

8.   Study Figure 4.12 in your text and explain the effects of hemicholinium.

9.   List two types of acetylcholine receptors and briefly describe their characteristics.

1.

2.

10. *Belladonna* means "pretty lady." Explain why atropine is referred to as a belladonna alkaloid.

11. Within minutes of receiving curare, what physical changes occur in the body? Explain how the drug works and describe its use in anesthesia.

Learning Objective 4-5   Describe the monoaminergic pathways in the brain and the drugs that affect these neurons.

Read pages 102-111 and answer the following questions.

1.   List and group by subclass the four transmitter substances that belong to the monoamine family. (See Table 4.3 in your text.)

2.   In general, how do the monoamines affect brain functions?

3.   What are some of the functions in which dopaminergic neurons play a role?

4. Study Figure 4.13 in your text and briefly summarize the synthesis of the catecholamines from tyrosine.

5. There are several systems of dopaminergic neurons in the brain, of which three are most important. Name them, describe where their axons project, and say what functions they affect.

    1.

    2.

    3.

6. Name and describe a serious disorder caused by the degeneration of dopaminergic neurons. What drug is effective in the treatment of this disorder and how does it work?

7. Briefly describe the activity of the drug AMPT.

8. What monoamine antagonist was discovered over three thousand years ago in India? How does it function?

9. Identify the two most common dopamine receptors and briefly describe their characteristics.

    1.

    2.

10. Study Figure 4.15 in your text and review the activity of the drug apomorphine.

    a. Apomorphine is a _____ _____, but seems to have a greater affinity for _____
       _____ _____ than _____ _____ _____.

    b. What are the effects of a low dose of apomorphine? a high dose?

11. Which drugs inhibit the reuptake of dopamine?

12. What is the role of monoamine oxidase (MAO) found in

    a. terminal buttons?

    b. blood?

13. Which drugs are used in the treatment of schizophrenia? What are their effects on dopamine receptors?

14. Adrenaline and _____ are synonymous, as are noradrenaline and _____.

15. a. Where does the final step in the synthesis of norepinephrine take place?

b. How is dopamine converted to norepinephrine?

c. What drug inhibits the activity of the enzyme dopamine β-hydroxylase, and what are its effects?

d. What does the drug moclobemide do?

16. a. What general regions of the brain receive input from noradrenergic neurons?

b. The cell bodies of the most important noradrenergic system begin in the _____
_____. (See Figure 4.17 in your text.)

c. What is the primary behavioral effect of activation of these neurons?

17. Through what structure do most neurons release norepinephrine? What is their appearance?

18. What are the effects of activation of α_1, α_2, and β adrenergic receptors?

19. Study Figure 4.18 in your text and briefly summarize the synthesis of serotonin (also called 5-HT, or 5-hydroxytryptophan) from tryptophan.

20. Where are the cell bodies of serotonergic neurons found? Which are the most important clusters?

21. The D system originates in the _____ raphe nucleus. Its axonal fibers are _____, with _____ _____. The M system originates in the _____ raphe nucleus. Its axonal fibers are _____ and _____.

22. How many types of serotonin receptors have been identified? What are some of their behavioral roles?

23. List three drugs that interact with serotonin and describe their effects.

    1.

    2.

    3.

Learning Objective 4-6   Review the role of neurons that release amino acid neurotransmitters and describe drugs that affect these neurons.

Read pages 111-115 and answer the following questions.

1. a. Approximately how many amino acids may serve as transmitter substances in the mammalian central nervous system?

   b. Name the three most common of them.

   1.   3.

   2.

   c. Name the two thought to have evolved first.

   1.   2.

2. List the four types of glutamate receptors.

   1.   3.

   2.   4.

3. a. Describe some special characteristics of the NMDA receptor.

   b. Why is the entry of calcium into the cell important? What are its effects?

   c. Study Figure 4.21 and identify the binding sites on the NMDA receptor.

4. a. Describe how PCP exerts its effects.

   b. List seven behavioral symptoms of PCP. (See Table 4.7 in your text.)

5. a. Explain how gamma-aminobutyric acid (GABA) is synthesized and describe where it is found.

    b. Identify the two GABA receptors and briefly describe their functions.

    c. Explain the general effect of GABA and indicate its importance in maintaining stability in the brain.

6. Briefly describe the effects of muscimol and bicuculline.

7. What class of drugs is used to relieve anxiety? On which site do these drugs bind?

8. What are the effects of barbiturates? Why are they a poor choice for anesthesia?

9. What are the effects of picrotoxin? On which site does it bind?

10. What chemical has tentatively been identified as an endogenous ligand for the benzodiazepine binding site? What are its behavioral effects and possible function?

11. a. Where is glycine found and what is its principal effect?

    b. Explain the effect of tetanus toxin on glycine synapses.

    c. What are the effects of strychnine?

Learning Objective 4-7   Describe the effects of peptides, lipids, nucleosides, and soluble gases released by neurons.

Read pages 115-118 and answer the following questions.

1. Peptides are chains of _____ _____ that are produced by precursor molecules, which are large _____. Peptides are synthesized in the _____ of neurons. After they are released, they are deactivated by _____ instead of being recycled.

2. Define *endogenous opioid.*

3. Describe several phenomena that are influenced by peptides.

4. How does naloxone function, and what are its practical uses?

5. What may be the role of peptides that are released along with a transmitter substance? Describe research to support your answer.

6. What is the role of lipids in neural communication? Describe research on the THC receptor to support your answer. (Matsuda et al., 1990; Devane et al., 1992)

7. A nucleoside is a compound that consists of a(n) _____ _____ bound with a(n) _____ or _____ base.

8. Describe adenosine, one of these compounds.

    a. What is its general function?

    b. Where and why is it released?

    c. How many types of receptors does it act upon?

    d. What are the physiological effects of activation of adenosine receptors?

    e. What common drug blocks adenosine receptors?

9. a. Name two soluble gases involved in neural communication.

    b. Which bodily functions are affected by nitric oxide? (Culotta and Koshland, 1992)

    c. Where is nitric oxide produced, how quickly does it diffuse, and how long is it present in nearby cells?

Lesson II Self Test

1. Acetylcholine

    a. has an excitatory effect on cardiac muscle fibers.
    b. is found in the autonomic nervous system.
    c. is deactivated by reuptake.
    d. produces hyperpolarization in skeletal muscle fibers.

2. Select the incorrect statement.

    a. Acetylcholine plays a role in attention and reinforcement.
    b. Acetylcholine is made of acetate and choline.

    c. Acetylcholine acts on both nicotinic and muscarinic receptors.
    d. Acetylcholine has both inhibitory and excitatory effects on cells.

3. _____ blocks muscarinic receptors and _____ blocks nicotinic receptors.

    a. Curare; atropine
    b. Apomorphine; physostigmine
    c. Atropine; curare
    d. Tyrosine; atropine

4. Which of the following is not a monoamine neurotransmitter?

a. tyrosine
b. serotonin
c. norepinephrine
d. dopamine

5. Which of the following statements is true?

a. Dopamine is synthesized from the amino acid tryptophan.
b. Dopamine is involved in arousal and sleep.
c. Dopamine stimulates at least five different kinds of ionotropic receptors.
d. Dopaminergic synapses are lacking in Parkinson's disease and overactive in schizophrenia.

6. The function of monoamine oxidase (MAO) is to

a. facilitate synthesis of the monoamines.
b. destroy excess amounts of the monoamines.
c. block the reuptake of the monoamines.
d. transport the monoamines down the axon to the terminal buttons.

7. Which drug was the first to receive widespread use as an effective treatment for the symptoms of schizophrenia?

a. Valium
b. chlorpromazine
c. atropine
d. lithium carbonate

8. Serotonergic neurons

a. are involved in the symptoms of schizophrenia.
b. help regulate blood pressure.
c. are involved in the control of dreaming.

d. are involved with balance and walking.

9. Glutamic acid, GABA, and glycine

a. may be the three most common transmitter substances in the CNS.
b. appeared late in the evolutionary process.
c. circulate through the brain in the cerebrospinal fluid.
d. are the principal excitatory transmitter substances of the brain.

10. All of the following act on the GABA receptor complex except

a. NMDA.
b. benzodiazepines.
c. barbiturates.
d. alcohol.

11. Many peptides that are released at the same time as a transmitter substance may serve to regulate the

a. sensitivity of postsynaptic receptors to the transmitter substance.
b. reuptake of the transmitter substance.
c. axoplasmic flow to the terminal buttons.
d. metabolism of the brain.

12. Endogenous opioids are

a. by-products of the synthesis of transmitter substances.
b. steroids synthesized from cholesterol.
c. neuromodulators produced in the brain.
d. hormones secreted by many tissues of the body.

Answers for Self Tests

Lesson I

1.	c	Obj. 4-1
2.	d	Obj. 4-1
3.	a	Obj. 4-2
4.	a	Obj. 4-2
5.	c	Obj. 4-2
6.	b	Obj. 4-3
7.	c	Obj. 4-3
8.	d	Obj. 4-3
9.	c	Obj. 4-3
10.	a	Obj. 4-3
11.	d	Obj. 4-3
12.	b	Obj. 4-3

Lesson II

1.	b	Obj. 4-4
2.	a	Obj. 4-4
3.	c	Obj. 4-4
4.	a	Obj. 4-5
5.	d	Obj. 4-5
6.	b	Obj. 4-5
7.	b	Obj. 4-5
8.	c	Obj. 4-5
9.	a	Obj. 4-6
10.	a	Obj. 4-6
11.	a	Obj. 4-7
12.	c	Obj. 4-7

Chapter 5
Methods and Strategies of Research

Lesson I: Experimental Ablation

Read the interim summary on pages 132-133 of your text to re-acquaint yourself with the material in this section.

Learning Objective 5-1   Discuss the research method of experimental ablation: the rationale, the evaluation of behavioral effects resulting from brain damage, and the production of brain lesions.

Read pages 120-123 and answer the following questions.

1.   Define *experimental ablation* in your own words.

2.   Now define *lesion study* and explain its rationale.

3.   a. Distinguish between *brain function* and *behavior* and explain why this distinction is important in interpreting research results.

   b. What anatomical difficulty do researchers encounter in interpreting results of lesion studies?

4.   Briefly describe and compare research techniques to produce brain lesions.

   a. Suction

      1. procedure

      2. selectivity

   b. Radio frequency (RF) current (See Figure 5.1 in your text.)

      1. procedure

      2. selectivity

   c. Excitatory amino acids such as kainic acid that produce excitotoxic lesions. (See Figure 5.2 in your text.)

      1. procedure

2. selectivity

d. Drugs such as 6-hydroxydopamine (6-HD)

1. procedure

2. selectivity

5. Whenever subcortical regions are destroyed, what other kind of damage is unavoidable?

6. What procedure do researchers follow to try to determine whether this damage has affected their results? (Be sure to use the term *sham lesion* in your answer.)

7. Now briefly describe how brain activity can be temporarily interrupted using local anesthetics or cooling techniques. (See Figure 5.3 in your text.)

Learning Objective 5-2   Describe stereotaxic surgery.

Read pages 123-125 and answer the following questions.

1. Name the instrument and the reference book researchers use during stereotaxic surgery.

2. a. Describe the contents and organization of a stereotaxic atlas.

b. Now describe how a researcher uses the atlas to locate a subcortical brain structure. Be sure to use the term *bregma* in your answer. (See Figures 5.4 and 5.5 in your text.)

c. Why are the locations described in stereotaxic atlases only approximate?

3. Study Figure 5.6 in your text and describe a stereotaxic apparatus.

4. How is an animal prepared and positioned for surgery?

5. List several uses of stereotaxic surgery.

Learning Objective 5-3   Describe research methods for preserving, sectioning, and staining the brain and for studying its parts and interconnections.

Read pages 125-127 and answer the following questions.

1. Why do researchers use histological methods to prepare brain tissue for microscopic examination?

2. After the brain is removed from the skull, it is placed in a _____ such as _____, which _____ the tissue, halts _____ and kills any _____ that might destroy it.

3. Why is tissue usually perfused before placing it in a fixative?

4. Study Figure 5.7 in your text and describe a microtome and its function.

5. Briefly explain how neural tissue is prepared for staining.

6. Explain why brain tissue must be stained.

7. If a researcher stains neural tissue with cresyl violet, what kind of material will take up the stain? (See Figure 5.8 in your text.)

8. a. When do researchers use an electron microscope rather than a light microscope?

    b. In general, how do electron microscopes and scanning electron microscopes produce images?

    c. Study Figure 5.9 and 5.10 in your text and compare those images.

Learning Objective 5-4   Describe research methods for tracing efferent and afferent axons and for studying the living human brain.

Read pages 127-132 and answer the following questions.

1. Sometimes a researcher may wish to trace pathways of *efferent* axons—that is, those that leave a particular brain structure—in order to learn more about how that structure may ultimately influence behavior.

a. What is the general name for all techniques used for this purpose?

b. Write a description of the method that uses PHA-L using the following phrases, some of which are shown in Figures 5.12 in your text.

travel by means of fast axoplasmic transport   cells are filled with molecules of PHA-L

into the brain structure being studied   slice and mount the brain tissue

examine tissue under the microscope   use an immunocytochemical method

molecules of PHA-L are taken up by dendrites   inject a minute quantity of PHA-L

to the terminal buttons   kill the animal

transported through the soma to the axon   within a few days

c. Carefully explain how immunocytochemical methods make use of our knowledge of the role of proteins, antibodies, and antigens in the immune system.

2. Researchers must also trace pathways of *afferent* axons; that is, those that lead to a particular brain structure.

a. What is the general name for all techniques used for this purpose?

b. Name a chemical used to trace afferent axons and the means by which it is transported to the cell bodies. Some results are shown in Figures 5.14 and 5.15 in your text.

3. How has the development of computerized tomography (CT) and magnetic resonance imaging (MRI) overcome some of the earlier difficulties inherent in studying the human brain? (See Figures 5.16 in your text.)

4. Compare and contrast computerized tomography and MRI. (An example of a CT scan in shown in Figure 5.17 in your text and an example of a MRI scan is shown in Figure 5.18.)

a. procedure used to obtain a scan

b. details shown in scans

To review all the research methods discussed in this section, study Table 5.1 in your text.

Lesson I Self Test

1. The results of lesion studies are often difficult to interpret because

   a. a particular neural circuit can perform only one behavior.
   b. directly or indirectly, all regions of the brain are interconnected.
   c. sometimes there is no clear difference between behavior and function.
   d. the effects of surgical anesthesia cannot be determined.

2. Which of these methods produces the most selective brain lesions?

   a. suction
   b. microdialysis
   c. excitatory amino acids
   d. radio frequency current

3. To account for incidental brain damage when lesions are produced, researchers

   a. increase the number of animals in the study.
   b. use equal numbers of male and female animals.
   c. produce sham lesions.
   d. repeat the study.

4. A temporary or "reversible" brain lesion can be produced using

   a. very low doses of an excitatory amino acid.
   b. a local anesthetic.
   c. electrical current on only one side of the brain.
   d. a very brief interruption in the blood supply.

5. Using a stereotaxic apparatus, researchers can

   a. assess loss of function resulting from brain lesions.
   b. slice brains for histological examination.
   c. make subcortical lesions in the brain.
   d. confirm the location of brain lesions.

6. Using a stereotaxic atlas, researchers can

   a. locate bregma.
   b. determine the approximate location of structures deep within the brain.
   c. calculate the correct amount of anesthesia required.

   d. determine which method of making brain lesions to use.

7. A fixative performs all these functions except

   a. attaching tissue to a microscope slide.
   b. halting autolysis.
   c. hardening tissue.
   d. killing destructive microorganisms.

8. Using a microtome, researchers can

   a. apply a mounting medium.
   b. dry and heat tissue.
   c. examine and photograph stained and mounted brain sections.
   d. slice tissue into thin sections.

9. Scanning electron microscopes produce

   a. moving images that can be preserved on video cassettes.
   b. images cast on glass slides that can also be examined with a light microscope.
   c. images of three dimensional structures.
   d. images of tissue that cannot be exposed to light.

10. _____ labeling methods are used to trace _____ axons which carry information _____ a brain structure.

    a. Retrograde; afferent; toward
    b. Anterograde; afferent; away from
    c. Retrograde; efferent; away from
    d. Anterograde; afferent; toward

11. Immunocytochemical methods are used to

    a. locate peptides and proteins.
    b. provide a two-dimensional view of the human brain.
    c. selectively destroy axons.
    d. inhibit fast axoplasmic transport.

12. Researchers using CT or MRI

    a. do not have to obtain permission from either the patient or the family.
    b. can study the living brain without operating on the patient.
    c. anesthetize the patient before beginning.
    d. must first shave the patient's head.

Lesson II: Recording and Stimulating Neural Activity and Neurochemical Methods

Read the interim summary on pages 141-142 of your text to re-acquaint yourself with the material in this section.

> *Learning Objective 5-5*   Describe how the neural activity of the brain is measured and recorded, both electrically and chemically.

Read pages 134-139 and answer the following questions.

1. To review: List the two types of electrical events in the brain that can be recorded.

    1.   2.

2. Compare the duration and behavioral state of the animal during chronic and acute recordings.

3. Explain what the phrase *single-unit recording* means.

4. Describe the preparation for recording the neural activity of individual neurons in the brain.

    a. How are microelectrodes of glass and tungsten wire made? (See Figure 5.19 in your text.)

    b. How are these electrodes implanted in the brain? (See Figure 5.20 in your text.)

    c. How are the electrical signals detected by the microelectrodes recorded?

5. How does the preparation change for recording the neural activity of a whole region of the brain? Be sure to describe macroelectrodes in your answer.

6. a. Describe two ways to measure the electrical activity of the human brain.

    b. By studying these records of electrical patterns, what can researchers learn? (See Figures 5.21 and 5.22 in your text.)

7. a. In addition to the electrical activity of the brain, what other sign of neural activity can be measured?

b. Explain the role of 2-deoxyglucose (2-DG) in measuring the metabolic activity of the brain.

c. Explain how autoradiographs are prepared. (See Figure 5.23 in your text.)

8. a. How are nuclear proteins produced and what do they indicate?

   b. Describe the procedure that detects the presence of Fos and explain what is shown in Figure 5.24 in your text.

9. a. Explain how the metabolic activity of the human brain is measured using positron emission tomography (PET). (See Figure 5.25 in your text.)

   b. What is one of the reasons why PET scanners are expensive to use?

10. Explain several advantages of functional magnetic resonance imaging (fMRI). (See Figure 5.26 in your text.)

11. Study Figure 5.27 in your text and then carefully explain how neurochemicals can be detected using microdialysis.

12. How can neurochemicals be measured in the human brain by means of a noninvasive technique? (See Figure 5.28 in your text)

Learning Objective 5-6   Describe how neural activity in the brain is stimulated, both chemically and electrically and the behavioral effects of electrical brain stimulation.

Read pages 139-141 and answer the following questions.

1. List two ways that neurons can be artificially stimulated.

    1.   2.

2. Briefly explain both procedures, mentioning any advantages and disadvantages. (See Figure 5.29 in your text.)

3. a. Describe how extremely small quantities of a substance are discharged through a multibarreled micropipette which is illustrated in Figure 5.30 in your text.

    b. How is neural activity recorded during microiontophoresis?

4. List three uses of electrical brain stimulation.

    1.   3.

    2.

5. Identify and briefly discuss some of the difficulties in interpreting the significance of behavioral changes elicited by electrical brain stimulation.

6. a. Why are patients undergoing open-head surgery to remove a seizure focus not given a general anesthetic?

    b. How do surgeons determine the function of neural tissue near the seizure focus?

    c. How did Penfield, who developed this treatment for focal seizure disorder, use the data he obtained using electrical brain stimulation? (See Figure 5.31 in your text.)

To review all the research methods discussed in this section, study Table 5.2 in your text.

Read the interim summary on page 146 of your text to re-acquaint yourself with the material in this section

---

Learning Objective 5-7   Describe research methods for locating particular neurochemicals, the neurons that produce them, and the receptors that respond to them.

Read pages 142-146 and answer the following questions.

1. Let's follow the example of the effects of organophosphate insecticides to learn how researchers look for a particular neurochemical and, at the same time, review information about acetylcholine.

    a. State the two hypotheses that could explain why organophosphate insecticides disrupt dreaming.

    b. Which one of these is correct?

c. Carefully explain how acetylcholinesterase inhibitors such as the organophosphate insecticides affect the brain.

d. Further research is still needed to determine in which region of the brain the acetylcholinergic synapses are affected. Outline three possible strategies to search for the affected region.

2. Now outline three ways to search for a particular neurochemical.

3. a. Briefly explain how brain tissue is prepared using immunocytochemical methods. (See Figure 5.32 in your text.)

b. Which one of the ways outlined in question 2 can be studied using immunocytochemical methods?

c. Name the enzyme involved in the synthesis of acetylcholine and explain how it is identified using immunocytochemical methods. (See Figure 5.33 in your text.)

4. a. Which one of the ways outlined in question 2 can be studied using in situ hybridization?

b. Study Figures 5.34 and 5.35 in your text and briefly explain how in situ hybridization is used to find the location of a particular messenger RNA.

c. In addition to location, what other information does this method provide?

5. List two methods and briefly explain how they are used to search for acetylcholine receptors.

6. Explain how we can verify that the VMH is affected by ovarian sex hormones by using autoradiography and immunocytochemistry.

7. What kinds of information can be obtained using double labeling? (See Figure 5.36 in your text.)

To review all the research methods discussed in this section, study Table 5.3 in your text.

Lesson II Self Test

1. Microelectrodes are used to record the electrical activity of

    a. individual neurons in the brain.
    b. entire regions deep within the brain.
    c. the fluid that fills the ventricles.
    d. the synaptic cleft.

2. 2-deoxyglucose is

    a. found in the membrane of cells.
    b. a nuclear protein.
    c. the preferred fuel of the brain.
    d. not metabolized by cells.

3. An electroencephalogram is a

    a. a microscope slide of radioactivity in the brain.
    b. paper record of electrical activity in the brain.
    c. x-ray record of normal or abnormal brain tissue.
    d. treated photographic film showing receptor location.

4. The metabolic activity in the brain of a laboratory animal can be measured by giving an injection of _____ and analyzing the results using _____.

    a. Fos; PET scanner
    b. 2-DG; autoradiography
    c. fluorogold; high-performance liquid chromatography
    d. saline; electroencephalography

5. Functional MRI (fMRI)

    a. was the first imaging technique for studying the living brain.
    b. images were never as detailed as those obtained using PET scanners.
    c. permits the measuring of regional metabolism.
    d. can be used to locate any radioactive substance.

6. Microdialysis is used to measure the

    a. permeability of the cell membrane.
    b. intracranial pressure.
    c. secretions of the brain.

    d. electrical activity of the brain.

7. An advantage of chemical stimulation over electrical stimulation of the brain is that

    a. it requires less equipment.
    b. stereotaxic surgery is not necessary.
    c. the effects are more localized.
    d. no tissue is destroyed.

8. To study the effect of chemicals on the activity of a single cell researchers use

    a. microiontophoresis.
    b. in situ hybridization.
    c. single-unit recording.
    d. autoradiography.

9. One of the difficulties in using electrical brain stimulation as a research tool is stimulation

    a. sometimes triggers a seizure.
    b. is difficult to localize.
    c. must always be administered during open head procedures.
    d. can never duplicate natural neural processes.

10 To identify neurons producing a particular peptide, researchers use _____ methods.

    a. immunocytochemical
    b. anterograde tracing
    c. autoradiographic
    d. retrograde tracing

11. To search for a particular messenger RNA, researchers use

    a. the Fos protein.
    b. autoradiography.
    c. in situ hybridization.
    d. 2-DG.

12. Double labeling techniques permit researchers to determine what chemicals a neuron contains and

    a. their connections with other neurons.

b. the enzymes that produce them.

c. the messenger RNA involved in their synthesis.

d. nearby agonists or antagonists.

Answers for Self Tests

Lesson I			Lesson II		
1.	b	Obj. 5-1	1.	a	Obj. 5-5
2.	c	Obj. 5-1	2.	d	Obj. 5-5
3.	c	Obj. 5-1	3.	b	Obj. 5-5
4.	b	Obj. 5-1	4.	b	Obj. 5-5
5.	c	Obj. 5-2	5.	c	Obj. 5-5
6.	b	Obj. 5-2	6.	c	Obj. 5-5
7.	a	Obj. 5-3	7.	c	Obj. 5-6
8.	d	Obj. 5-3	8.	a	Obj. 5-6
9.	c	Obj. 5-3	9.	d	Obj. 5-6
10.	a	Obj. 5-4	10.	a	Obj. 5-7
11.	a	Obj. 5-4	11.	c	Obj. 5-7
12.	b	Obj. 5-4	12.	a	Obj. 5-7

CHAPTER 6
Vision

Lesson I: Anatomy of the Visual System and Coding of Visual Information

Read the interim summary on page 157 of your text to re-acquaint yourself with the material in this section.

> *Learning Objective 6-1*   Describe the characteristics of light and color, outline the anatomy of the eye and its connections with the brain, and describe the process of transduction of visual information.

Read pages 149-157 and answer the following questions.

1. We receive information about the environment from sensory receptors. What is the name of the process by which stimuli impinge on the receptors? What is the name of the electrical change in the cells' membrane?

    transduction

2. _____ is that portion of the electromagnetic spectrum that we humans can see. Light travels at a

    constant speed of _____ per second and the wavelength of visible light determines one of the

    perceptual dimensions of color, the _____. When the intensity of the electromagnetic signal

    increases, we perceive that the _____ of the color has increased. The degree of _____ of

    a color depends on the purity of the light. (See Figures 6.1 and 6.2 in your text.)

3. List the three types of movement that eyes make and describe them briefly.

    1. *vergence — movement from near to far*

    2. *saccade — rapid, involuntary back + forth movements*

    3. *smooth pursuit — slow, conscious control*

4. Begin a review of the anatomy of the eye by studying Figures 6.3 and 6.4 in your text.
    a. Name the bony pockets in the front of the skull that surround the eyes.   *ocular orbits*
    b. What holds the eyes in place and moves them?
    c. Name the white outer layer of the eye.   *cornea*
    d. Light enters through which layer of the eye?   *cornea*
    e. Describe how the amount of light that enters the eye is regulated.

f. Describe the lens and explain the process of accommodation.

g. Name the substance between the lens and retina that gives the eye its volume.   *vitreous humor*

h. Name the light-sensitive region of the back of the eye.   *fovea*

5. List the characteristics of the specialized photoreceptors—the rods and the cones—in Table 1, below.

	Cones	Rods
Number	*More*	*less*
Location	*Fovea*	*Periphery*
Sensitive to what kind of light?	*High*	*low*
Role in visual acuity		
Role in color vision	*Color*	*B/W*

Table 1

6. Label the parts of the eye in Figure 1, below.

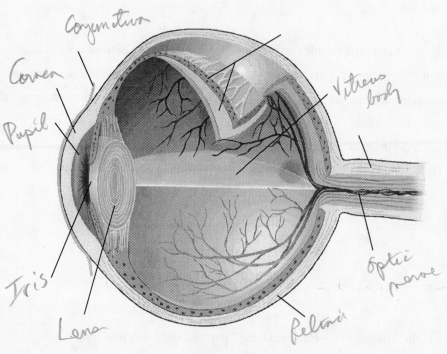

Conjunctiva
Cornea
Pupil
Iris
Lens
Vitreous body
Optic nerve
Retina

Figure 1

7. Explain why the optic disk is called the blind spot. Be sure to try the demonstration in Figure 6.5 in your text.

8. a. Study Figure 6.6 in your text and list the three primary layers of the retina.

1.   3.

2.

b. Where are the photoreceptors located in relation to these layers and to the direction of entering light?

c. Photoreceptors form synapses with _bipolar cells_, which form synapses with _ganglion cells_.

d. Name two other kinds of cells found in the retina.

1.   2.

9. Study Figure 6.7 in your text for a closer look at the anatomy of photoreceptors. Sketch a rod and a cone and label their parts.

10. Now let's follow the process of visual perception step by step. We will begin with the role of the photopigments.

a. List the two components of photopigments.

1.   2.

b. Indicate on your sketch in question 9 where photopigments are found.

c. Name the photopigment found in human rods.

d. Describe how rhodopsin changes when it is exposed to light.

e. List two ways in which photoreceptors differ from other neurons.

1.   2.

f. Study Figure 6.8 in your text and carefully explain the changes that occur in the photoreceptor when it is struck by light. Be sure to use the term *transducin* in your answer.

11. Now let's continue by studying the neural circuit from the photoreceptors to the ganglion cells.

a. Study Figure 6.9 in your text and list the first two types of cells in this circuit.

1.   2.

b. Light _hyper_polarizes the photoreceptors and _de_polarizes the bipolar cells.

c. What happens to the firing rate of this particular ganglion cell?

12. Axons of the retinal ganglion cell ascend through the optic nerves to which part of the brain?

lateral geniculate nucleus

13. a. What is the name of the inner two layers of the lateral geniculate nucleus? the outer four layers? (See Figure 6.10 in your text.) *Magnocellular neurons*

   Parvocellular

b. What do these names indicate about the size of the cells found there?

   Magno = big   parvo = small

14. a. To what region of the cerebral cortex do axons of neurons in the dorsal lateral geniculate nucleus project?

b. Where is this region located? *striate cortex , occipital lobe*

c. Study Figure 6.11 in your text and explain why the primary visual cortex is sometimes called the striate cortex.

15. Study Figure 6.12 in your text to review the primary visual pathway in more detail. Trace the route of the optic nerves from the optic chiasm to the dorsal lateral geniculate nucleus and explain why the optic chiasm is so named.

16. a. If a person looks straight ahead, the right hemisphere receives information from the _____ side of the visual field and the left hemisphere receives information from the _____.

b. Explain why the previous statement is true. Be sure to explain the role of the lens in your answer.

17. Briefly describe the functions of several other pathways from the retina.

Read the interim summary on pages 163-164 of your text to re-acquaint yourself with the material in this section.

Learning Objective 6-2   Describe the coding of visual information by photoreceptors and ganglion cells in the retina.

Read pages 157-164 and answer the following questions.

1. a. Define the *receptive field* of a neuron in your own words.

b. Using the concept of receptive fields, explain why foveal vision is very accurate. (See Figure 6.13 in your text.)

2. List and describe

   a. the response pattern of ganglion cells in the frog retina, discovered by Hartline (1938).

   b. the response pattern of ganglion cells in the cat retina, discovered by Kuffler (1952, 1953). (See Figure 6.14 in your text.)

   c. the connections of ON/OFF cells in the primate retina. (Schiller and Malpeli, 1977)

3. Why could the firing patterns of ON and OFF cells be described as efficient?

4. a. What kind of information is detected by the ON and OFF cells in the monkey retina? (Schiller et al., 1986)

   b. Explain how we know that rod bipolar cells in the monkey retina must be ON cells. (Dolan and Schiller, 1989)

5. a. Describe a phenomenon that results from the center-surround organization of the receptive fields of ganglion cells. (Look at Figure 6.15 for an example.)

   b. Study Figure 6.16 and explain how this phenomenon works.

6. State two advantages of color vision. (Mollon, 1989)

7. Briefly state Thomas Young's theory of color vision.

8. Before continuing, study Figure 6.17 in your text and explain the difference between color mixing and pigment mixing.

9. Briefly state Ewald Hering's theory of color vision.

10. What two problems with the Young theory did Hering's theory overcome?

    1.

    2.

11. a. Later research with photoreceptors in the retina proved Young was right. Study Figure 6.18 in your text and explain the relationship between the particular opsins found in retinal photoreceptors and color vision.

    b. List the three types of cones found in the retina and compare their relative numbers.

    1.   3.

    2

12. Supply characteristics for these three genetic defects in color vision.

Defect	Description	Visual Acuity	Sex ratio
Protanopia			
Deuteranopia			
Tritanopia			

13. Retinal ganglion cells respond to _____ of opposing primary colors, thus there are two kinds of color-sensitive cells _____-_____ and _____-_____.

14. Describe the typical response pattern of color-sensitive and "black and white" retinal ganglion cells. (Study Figure 6.19 in your text.)

15. Study Figure 6.20 in your text and describe how retinal ganglion cells code different wavelengths of light detected by the cones.

    a. How does red light affect

    1. red cones?   2. red-green ganglion cells?

    b. How does green light affect

    1. green cones?   2. red-green ganglion cells?

    c. How does yellow light affect

    1. red and green cones?   3. red-green ganglion cells?

    2. yellow-blue ganglion cells?

16. Now explain why we can imagine a yellowish red but not a bluish yellow.

17. Explain why you will see a red and green radish on the right of Figure 6.21 after staring at the left side for thirty seconds.

Read the interim summary on pages 171-172 of your text to re-acquaint yourself with the material in this section.

Learning Objective 6-3   Discuss the striate cortex and discuss how its neurons respond to orientation, movement, and spatial frequency.

Read pages 164-168 and answer the following questions.

1. Look back at Figure 6.11 and at Figure 6.22 and review the appearance of striate cortex.

2. What nucleus of the thalamus sends visual information to the middle layer?

3. Approximately what percentage of the striate cortex analyzes information from the fovea?

4. What discovery by Hubel and Wiesel (1977, 1979) revolutionized the study of the physiology of visual perception?

5. Study Figure 6.23 in your text and explain what is meant when we say, "Most neurons in the striate cortex are sensitive to orientation."

6. Study Figure 6.24 in your text and compare the responses of simple and complex cells to various line orientations and backgrounds as well as movement.

7. Explain the importance of sine-wave gratings in visual detection (De Valois et al., 1978).

8. Study Figure 6.25 in your text and use a pencil to sketch and shade a square-wave grating and a sine-wave grating.

9. a. Study Figure 6.26 in your text and explain the concept of visual angle.

   b. Go on to explain its relationship to the spatial frequency of sine-wave gratings.

10. Summarize research on the receptive fields of simple cells by Albrecht (1978).

   a. Describe or sketch the stimulus. (Study Figure 6.27 in your text.)

   b. How did the cells respond to the moving stimulus?

   c. What did the response pattern resemble?

11. Compare the spatial frequencies that indicate small objects or large objects with sharp edges and large areas of light and dark.

12. a. The most important visual information for object recognition is contained in _____ spatial frequencies.

   b. Compare the photographs of Abraham Lincoln in Figure 6.28 in your text and explain how the figure on the left was created. (Harmon and Julesz, 1973) You may want to try the demonstration as you answer this question.

   c. Explain how these figures confirm the statement in question 12a.

13. a. Describe the new class of neurons in the monkey striate cortex that von der Heydt et al. (1992) discovered.

   b. Approximately how many of these cells do they estimate there are in the monkey striate cortex?

   c. What useful function do they probably perform?

Learning Objective 6-4   Discuss how neurons in the striate cortex respond to retinal disparity and color; explain the modular organization of striate cortex and the phenomenon of blindsight.

Read pages 167-171 and answer the following questions.

1. a. List some of the cues we can observe using only one eye that help us determine depth.

b. Describe how we also determine depth through stereopsis using cues we observe with both eyes. Be sure to explain the concept of retinal disparity in your answer.

2. a. What observation led Wong-Riley (1978) to discover blobs?

b. Within striate cortex, where are blobs found? See Figure 6.31 and describe their appearance.

3. The brain is most likely organized in _____. The striate cortex consists of approximately 2500 _____, each containing about _____ neurons. These modules consist of two segments each centered around a _____.

4. a. How do the neurons within a blob differ from other neurons outside the blob? (Livingstone and Hubel, 1984; Edwards et al. 1995)

b. What do neurons outside the blob respond to?

5. Each half of the module receives input from only one eye. Why is information from these two inputs undoubtedly combined in striate cortex? (See Figure 6.32 in your text.)

6. Review Blasdel's (1992a, 1992b) research on the organization of modules in monkeys.

a. How was he able to see which neurons were excited in the striate cortex?

b. Study Figure 6.33a and describe his findings with regard to orientation sensitivity.

c. What does Figure 6.33b indicate about the relationship between ocular dominance and orientation sensitivity?

7. Refer to Figure 6.34 and describe how spatial frequency fits into the organization of CO blobs. (Edwards et al., 1995)

8. a. People with blindsight have damage to what part of the brain?

b. If someone has blindsight, what unexpected behavior will he or she perform? (Weiskrantz et al., 1974)

c   What is a possible explanation for blindsight?

Lesson I Self Test

1. If a color is fully saturated, the radiation contains

   a. all wavelengths.
   b. one wavelength.
   c. wavelengths beyond the visible spectrum.
   d. a mixture of wavelengths from a specific band of the spectrum.

2. Optic nerves join together at the _____, where half of the axons cross to the opposite side of the brain.

   a. calcarine fissure
   b. optic chiasm
   c. optic disk
   d. striate cortex

3. The first step in visual perception occurs when light

   a. causes a photopigment to split into its two constituents.
   b. enters the sclera.
   c. reaches the brain through the optic chiasm.
   d. causes a change in the receptor potential of the photoreceptor.

4. Foveal vision is more acute than peripheral vision because

   a. the receptor-to-axon relationships are approximately equal in the fovea.
   b. its photoreceptors respond more quickly to changes in illumination.
   c. its receptive field is near the fixation point.
   d. its ganglion cells fire continuously.

5. Kuffler found that the receptive field of cat ganglion cells resembles

   a. a mosaic.
   b. a circle surrounded by a ring.
   c. staggered columns.
   d. blobs.

6. Retinal ganglion cells use a(n) _____ coding system.

   a. trichromatic
   b. relative brightness
   c. opponent-process

   d. black and white

7. Hubel and Wiesel first suggested that orientation-sensitive neurons in the visual cortex responds best to _____, but other research indicates the best stimulus is _____.

   a. lines and edges; spots
   b. low spatial frequencies; high spatial frequencies.
   c. lines and edges; sine-wave gratings
   d. sine-wave gratings; spots

8. When low frequency information is removed from an image of an object, it becomes

   a. easier to identify.
   b. more difficult to identify.
   c. easier to identify if it is moved closer.
   d. more difficult to identify if it is moved closer.

9. Neurons that respond to "periodic patterns" are probably used to perceive

   a. surface texture.
   b. depth.
   c. color.
   d. high contrast images.

10. CO blobs

    a. were discovered in the modules of prestriate cortex.
    b. are organized in ovals within a module.
    c. contain color sensitive neurons.
    d. analyze information from the entire visual scene.

11. Retinal disparity helps us to recognize

    a. shapes and patterns.
    b. depth.
    c. density.
    d. right and left.

12. The phenomenon of blindsight includes all of the following except:

    a. the ability to accurately reach for an object held in one's blind field.

b. visual information that controls behavior without producing a conscious sensation.

c. connections that the visual association cortex receives from the superior colliculus and dorsal lateral geniculate nucleus.

d. an early evolutionary development.

Lesson II: Analysis of Visual Information

Read the interim summary on pages 183-184 of your text to re-acquaint yourself with the material in this section.

Learning Objective 6-5   Describe the anatomy of the visual association cortex and discuss the location and functions of the two streams of visual analysis that take place there.

Read pages 172-173 and answer the following questions.

1. a. Both streams of visual analysis of visual association cortex begin in the _____ _____,

     but begin to diverge in the _____ _____.

    b. Where does the ventral stream that turns downward end? (Study Figure 6.35 in your text.) What is its function?

    c. Where does the dorsal stream that turns upward end? What is its function?

2. Summarize the characteristics and functions of the parvocellular and magnocellular systems of the visual system by completing Table 2, below. (Study Table 6.1 in your text.)

Property	Magnocellular Division	Parvocellular Division
Color	No Sensitive	Sensitive
Sensitivity to contrast		
Spatial resolution	yes	No
Temporal resolution		

Table 2

3. Which system is found in all mammals? only primates?

4. Where does the dorsal stream receive its information? the ventral stream? (Maunsell, 1992)

5. a. Describe the location of extrastriate cortex. (Zeki and Shipp, 1988)

b. Briefly describe the organization of extrastriate cortex. Be sure to describe the flow of information.

Learning Objective 6-6   Discuss the perception of color and the analysis of form by neurons in the ventral stream.

Read pages 173-177 and answer the following questions.

1. Compare the response characteristics to color by neurons in the blobs and neurons in subarea V4 of extrastriate cortex. (Zeki, 1980)

2. Describe the phenomenon of color constancy in your own words.

3. Now describe research on color constancy by Schein and Desimone (1990).

    a. How did stimuli in the secondary receptive field influence the response to stimuli in the primary receptive field of neurons in V4?

    b. What do these responses suggest about color constancy?

4. How did Walsh et al. (1993) confirm the above findings?

5. What region did Heywood, Gaffan, and Cowey (1995) find, and what is its role?

6. Describe the vision of people with achromatopsia.

7. a. Describe the experimental procedure that Zeki et al. (1991) used to study the perception of color in humans.

    b. What changes were recorded by the PET scanner?

8. Explain why the perception of color and shape is interrelated.

9. a. Where does the analysis of form begin?

   b. Which regions of extrastriate cortex play a role?

   c. Where is the analyzed information sent?

10. a. Where is inferior temporal cortex located?

    b. List the two major regions of the inferior temporal cortex. (See Figures 6.36 and 6.37 in your text.)
       1.   2.

11. Describe these characteristics of neurons in area TEO.

    a. size of receptive field (Boussaoud et al., 1991)

    b. inputs

    c. outputs

    d. effects of lesions (For example, Iwai and Mishkin, 1969)

12. Now describe these characteristics of neurons in area TE.

    a. size of receptive field

    b. stimuli that evoke response

13. a. Describe the procedure Tanaka and his colleagues used to determine the stimuli that produced the best response in neurons in area TE.

    b. How did they change the stimulus during the experiment? (See Figure 6.38 in your text.)

c. What do the results suggest about the way neurons may perceive particular objects?

14. State the results of other research on responses of cells in the TE.

   a. Research displayed in Figure 6.39 in your text

   b. Research by Desimone et al. (1984)

15. Study Figure 6.40 in your text and describe what Wang et al. (1996) found about the activity of neurons in response to the turning of a doll's head.

16. What can we plausibly conclude about the neurons of the primate inferior temporal cortex?

Learning Objective 6-7   Describe the two basic forms of visual agnosia: apperceptive agnosia and associative visual agnosia.

Read pages 177-179 and answer the following questions.

1. a. Define *agnosia* in your own words.

    b. List the two forms of visual agnosia.

    c. Would you expect a person with visual agnosia to be able to
       1. read the list of contents of a first aid kit?
       2. identify by sight an adhesive bandage from among the other items in the kit?
       3. identify an adhesive bandage by touch?

    d. Explain why visual agnosia is a deficit in perception and identification rather than simple vision or memory.

2. Why would it be inappropriate to give a patient with prosopagnosia the following directions: "Look for John. He'll give you a ride home."

3. Which disorder indicates more severe brain damage, agnosia for common objects or prosopagnosia? (Alexander and Albert, 1983)

4. Describe how a patient studied by Damasio et al. (1982) found her car.

5. Which hemisphere is most important in the perception of faces?

6. a. Check the tasks that you would expect a patient with associative visual agnosia to be able to perform successfully. (Look at Figure 6.41 in your text.)

   _____ draw a picture of a tree

   _____ copy a picture of a tree

   _____ explain what a tree is

   b. Now explain why a patient can perform some, but not all of these tasks.

7. Explain how associative agnosia extends to prosopagnosia by citing research of Sergent and Signoret (1992).

8. Which disrupted brain connections appear to be responsible for associative visual agnosia?

Learning Objective 6-8   Describe how neurons in extrastriate cortex respond to movement and location and discuss the effects of brain damage on perception of these features.

Read pages 179-183 and answer the following questions.

1. a. Name and describe the location of neurons that respond to movement. (Look back at Figures 6.36 and 6.37.)

   b. List some of the regions that relay information to this area.

2. a. Circle the kinds of stimuli that elicit the best response from neurons in area V5. (Albright et al., 1984)

   1. moving stimuli / stationary stimuli

   2. movements in any direction / movements in a particular direction

   b. How is area V5 organized?

3. Review research by Salzman et al. (1992) on the perception of movement.

   a. What were the researchers measuring?

   b. What did the monkeys see? What were they supposed to do?

   c. What did the study demonstrate?

4. What may be the role of the pulvinar in the perception of movement? (Robinson and Petersen, 1992)

5. a   What kind of brain damage had the woman studied by Zihl et al. (1991) received?

    b. How was her perception of movement affected?

6. Describe the research performed by Malach et al. (1995) and their conclusions. (See Figure 6.42 in your text.)

7. a. Describe the technique Johansson (1973) and others used to study the kinds of information supplied by movement.

    b. What did subjects who saw the movies report?

8. Which regions may be damaged in people with visual agnosia who cannot recognize objects but can identify actions?

9. a. What area of the visual system projects to the parietal cortex? (Once again, look back at Figures 6.36 and 6.37.)

    b. What visual analysis is performed by the parietal lobes?

    c. What kind of deficits result from damage to this region? (Ungerleider and Mishkin, 1982)

10. a. What two discrimination tasks were studied by Haxby et al. (1994)?

    b. What relationship did they find between both discrimination tasks and metabolic activity? (See Figure 6.43 in your text.)

11. Describe the cause, symptoms, and resulting deficits of Balint's syndrome.
    a. People with Balint's syndrome have suffered brain damage to which part of the brain? (Balint, 1909; Damasio, 1985)

b. List and define the three major symptoms of this syndrome.

   1.

   2.

   3.

c. What is the characteristic response when a person with Balint's syndrome is asked to

   1. reach for an object?

   2. look around a room and describe its contents and their location? Why?

   3. identify several objects held together?

12. a. What terms do Goodale and colleagues prefer to use to describe the functions of the dorsal and ventral streams of visual cortex?

   b. Carefully state their reasons. (Cavada and Goldman-Rakic, 1989; Gentilucci and Rizzolatti, 1990)

   c. What kind of movements are impaired and remain intact as a result of brain lesions in this region? (Jakobson et al., 1991; Milner et al, 1991)

13. Explain why associative visual agnosia may result from disconnection of the ventral stream and verbal mechanisms in the brain. Cite research to support your answer. (Sirigu et al., 1991)

Lesson II Self Test

1. The ventral stream of visual association cortex recognizes

   a. the distance of an object from the viewer.
   b. where an object is located.
   c. the identity of an object.
   d. the movement of an object.

2. Extrastriate cortex

   a. consists of several regions that respond to a particular kind of visual information.
   b. receives information directly from the retina.
   c. performs the initial analysis of visual information.
   d. responds best to familiar stimuli.

3. Neurons in subarea V4 of extrastriate cortex

   a. have two receptive fields and respond to both.
   b. respond to a variety of wavelengths of light.
   c. are especially sensitive to changes in illumination.
   d. do not respond to orientation.

4. People with achromatopsia have

a. difficulty tracking a moving object.
b. lost some or all of their color vision.
c. no peripheral vision.
d. diminished visual acuity.

5. Inferior temporal cortex

   a. receives information from the magnocellular system.
   b. is found only in primates.
   c. performs the analysis of location and movement.
   d. performs the analysis of form and color.

6. All of the following are true of the TE area except:

   a. It is one of the two major regions of the inferior temporal cortex.
   b. Neurons of the TE have a large receptive field.
   c. Neurons in the TE respond well to simple stimuli such as spots, lines, or sine-wave gratings.
   d. Neurons in the TE appear to participate in the recognition of objects.

7. Associative visual agnosias are

   a. disconnections between visual perceptions and verbal systems.
   b. failures in high-level perceptions.
   c. genetic abnormalities involving the cones.
   d. inabilities to integrate tactile and visual information.

8. People with prosopagnosia

   a. have difficulty with visual accommodation.
   b. do not have binocular vision.

c. do not recognize faces.
d. are color blind.

9. Which of the following is not true of the V5 area?

   a. It is part of the extrastriate cortex, also known as MT.
   b. Neurons on the V5 respond to movement.
   c. The V5 is divided into rectangular modules.
   d. Most neurons of the V5 do not show directional sensitivity.

10. The pulvinar appears

   a. to monitor the position of the body in relation to the setting.
   b. to compensate for the effects of our own movements on movements of images on the retina.
   c. to coordinate eye and hand movements.
   d. to control eye fixation.

11. Simultanagnosia, one of the symptoms of Balint's syndrome, is the inability to

   a. reach for an object successfully.
   b. scan the contents of a room and perceive object locations.
   c. perceive more than one object in a group.
   d. identify objects by sight alone.

12. The dorsal stream may be primarily occupied with

   a. guiding actions.
   b. perceiving colors.
   c. identifying objects.
   d. controlling speech.

Answers for Self Tests

Lesson I		
1.	b	Obj. 6-1
2.	b	Obj. 6-1
3.	a	Obj. 6-1
4.	a	Obj. 6-2
5.	b	Obj. 6-2
6.	c	Obj. 6-2
7.	c	Obj. 6-3
8.	b	Obj. 6-3
9.	a	Obj. 6-3
10.	c	Obj. 6-4
11.	b	Obj. 6-4
12.	d	Obj. 6-4

Lesson II		
1.	c	Obj. 6-5
2.	a	Obj. 6-5
3.	b	Obj. 6-6
4.	b	Obj. 6-6
5.	d	Obj. 6-6
6.	c	Obj. 6-6
7.	a	Obj. 6-7
8.	c	Obj. 6-7
9.	d	Obj. 6-8
10.	b	Obj. 6-8
11.	c	Obj. 6-8
12.	a	Obj. 6-8

CHAPTER 7
Audition, the Body Senses, and the Chemical Senses

Lesson I: Audition and the Vestibular System

Read the interim summary on pages 202-203 of your text to re-acquaint yourself with the material in this section.

Learning Objective 7-1   Describe the parts of the ear and the auditory pathway.

Read pages 186-196 and answer the following questions.

1. a. What makes a sound? (See Figure 7.1 in your text.)

    b. What range of vibrations can humans hear?

2. Explain the following characteristics of sound. (See Figure 7.2 in your text.)

    a. pitch (Be sure to use the term *hertz* in your answer.)

    b. loudness

    c. timbre

3. Explain the difference between an analytic and a synthetic sense organ.

4. Study Figures 7.3 and 7.4 in your text and then describe the path of sound from your outer ear through your middle ear.

    a. When sound enters your external ear and moves through the external auditory canal what structure begins to vibrate?

    b. Describe the appearance and explain the function of the ossicles.

    c. Where is the oval window located?

5. Name the parts of the ear shown in Figure 1, below.

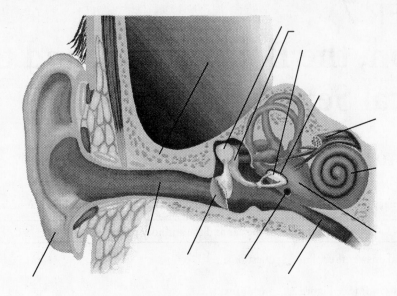

Figure 1

6. Briefly explain how airborne sound is transmitted to the cochlea.

7. a. List the names of the three sections of the cochlea shown in Figure 7.5 in your text.

    b. Name the three structures that compose the organ of Corti.

    c. Where are the hair cells found? What is their function?

    d. The cilia of the hair cells pass through the _____ _____, and some of them attach to the _____ _____.

8. a. Explain what von Békésy (1960) observed about the relationship between the portion of the basilar membrane that bent and the frequency of the sound.

    b. Study Figure 7.6 in your text and explain the role that the round window plays in hearing.

9. Describe the auditory hair cells.

    a. Where are the inner and outer auditory hair cells found?

    b. Describe their appearance and compare the number of each type. (See Figure 7.7 in your text.)

    c. What causes the basilar and tectorial membranes to bend and how do the cilia move in response?

10. Let's examine the role of cilia in more detail.

    a. Study Figures 7.8 and 7.9 in your text and explain how the direction of movement of the cilia affects their ion channels and results in an action potential.

    b. How did Denk and his colleagues (1995) discover the exact location of tip links? (See Figures 7.10 and 7.11.)

    c. Refer to Figure 7.12 and describe how most researchers think tension on tip links opens the ion channels.

11. a. Name the cranial nerve that connects the cochlea to the brain. Briefly describe its characteristics and components.

    b. Where do the vast majority of incoming cochlear nerve axons synapse? What do the connections of the inner hair cells suggest?

    c. What seems to be the function of the inner hair cells? the outer hair cells?

    d. Briefly describe the olivocochlear bundle.

12. Trace the central auditory pathway with the help of Figure 7.13 in your text. Begin with the cochlear nuclei of the medulla. Where do neurons of the

    a. cochlear nucleus send axons?

    b. superior olivary complex send axons?

    c. inferior colliculus send axons?

    d. medial geniculate nucleus send axons?

13. What other parts of the brain receive auditory information?

14. Describe the connections between the basilar membrane and the auditory cortex. How do we refer to the relationship between these two structures? Why?

Learning Objective 7-2   Describe the detection of pitch, loudness, timbre and the location of sound.

Read pages 196-202 and answer the following questions.

1. Explain the notion that moderate to high frequencies of sound are detected by place coding. Support your answer by referring to

   a. pioneering work of von Békésy.

   b. hearing loss induced by antibiotics. (Stebbins et al., 1969)

   c. cochlear implants.

2. Evans (1992) and Ruggero (1992) were able to determine something about the vibration of the basilar membrane that von Békésy was unable to discover. Describe their findings and what they suggest about the cells of the Organ of Corti.

3. Explain why the results obtained by Kiang (1965) suggested that low frequencies must be detected by some means other than place coding.

4. Describe the experiments by Pijl and Schwartz (1995a, 1995b) that provide evidence for rate coding.

5. Explain the means by which axons in the cochlear nerve inform the brain of the loudness of a stimulus. How is the loudness of low-frequency sounds signaled?

6. a. What characteristic of sound allows us to distinguish between a clarinet and a violin?

b. Study Figure 7.15 in your text and explain the mathematical analysis of a complex tone such as the sound of a clarinet. Be sure to use the terms *fundamental frequency* and *overtones* in your answer.

c. Explain how the basilar membrane responds to different overtones.

7. Name the two physiological mechanisms that we use to determine the source of a sound.

1.

2.

8. a. Imagine that you have spent the evening studying with a friend at her apartment. When she leaves you alone in the living room to go to the kitchen to make some tea, you notice the ticking of a clock. Explain how your eardrums will respond if the clock is

   1. on a table to the left.

   2. straight ahead of you.

   b. You hear the insistent bass notes of some music playing in an adjacent apartment. You ask your friend if she knows the person who lives in the apartment to the right, and whether she can ask the person to turn the music down. How did you know the music was coming from the right? Study Figure 7.16 in your text and be sure to use the terms *phase differences* and *out of phase* in your answer.

   c. What mechanism most likely detects short delays in the arrival times of signals? Explain how it works and cite supporting anatomical evidence (Carr and Konishi, 1989, 1990. See Figures 7.17 and 7.18 in your text.)

9. a. Why is the auditory system unable to use binaural phase differences to detect the location of high-frequency sounds?

   b. You hear the shrill whistle of your friend's tea kettle. You can hear that the sound comes from directly in front of you. Explain how you can perceive the direction from which it comes.

   c. Where are neurons that detect binaural differences in loudness found?

10. List the three primary functions of hearing. (Heffner and Heffner, 1990c; Yost, 1991)

    1.   3.

    2.

11. a. By what means does the auditory system appear to identify the sound source?

b. Where do the circuits of neurons involved in pattern recognition appear to be located?

c. Cite research to illustrate some of the kinds of analyses performed by auditory cortex. (For example, Whitfield and Evans, 1965)

12. What happens to the auditory abilities of monkeys following

a. bilateral lesions of the auditory cortex? How is their behavior affected? (Heffner and Heffner, 1990a)

b. lesions of the left auditory cortex? (Heffner and Heffner, 1990b)

Read the interim summary on page 205 of your text to re-acquaint yourself with the material in this section.

Learning Objective 7-3 Describe the structures and functions of the vestibular system.

Read pages 203-205 and answer the following questions.

1. a. List the two components of the vestibular system and the particular stimulus to which they respond.

   1.   2.

   b. List the functions of the vestibular system.

   1.   3.

   2.

2. List the two vestibular sacs.

   1.   2.

3. Study Figure 7.19 in your text and identify the bony labyrinths of the inner ear shown in Figure 2.

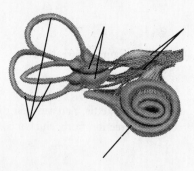

Figure 2

4. The semicircular canal consists of a _____ canal floating within a _____ one. An enlargement called the _____ contains the organ in which the sensory receptors reside. The cilia of

the sensory receptors are embedded in a gelatinous mass called the _____. (See Figure 7.20 in your text.)

5. Carefully describe how angular acceleration of the head affects the fluid in the semicircular canals.

6. Study Figure 7.21 in your text and explain how the vestibular sacs provide information on the head's orientation.

7. Study Figures 7.22 in your text and describe the appearance and function of the hair cells of the semicircular canal and the vestibular sacs.

8. List the two branches of the eighth cranial nerve–the auditory nerve.

    1.   2.

9. a. What is the location of the bipolar cells that give rise to the afferent axons of the vestibular nerve?

    b. Where do most afferent axons form synapses?

    c. Where do neurons of the vestibular nuclei send their axons?

10. Explain the importance of the vestibulo-ocular reflex.

Lesson I Self Test

1. The frequency of a vibration determines its

    a. pitch.
    b. loudness.
    c. timbre.
    d. complexity.

2. Because the cochlea is filled with fluid,

    a. its shape remains constant because liquids cannot be compressed.
    b. the ability to hear high frequencies diminishes with age as this liquid is absorbed by the body.

    c. sounds transmitted through air must be transferred to a liquid medium.
    d. it is most sensitive to rolling movements; thus we can experience seasickness.

3. Which of the following is not true about auditory hair cells?

    a. The inner hair cells are more numerous than the outer hair cells.
    b. They contain cilia.
    c. They produce depolarizations when bent in one direction and hyperpolarizations when bent in the opposite direction.

d. They contain actin filaments which make them stiff and rigid.

4. The round window moves in and out in opposition to movements of the

   a. Eustachian tube.
   b. basilar membrane.
   c. tectorial membrane.
   d. oval window.

5. Most of the neurons in the cochlear nuclei send axons directly to the

   a. auditory cortex.
   b. superior olivary complex.
   c. thalamus
   d. medial geniculate nucleus.

6. Progressive hair cell damage from the use of certain antibiotics causes a parallel progressive hearing loss, which suggests that some sounds are detected through

   a. place coding.
   b. rate coding.
   c. synthetic coding.
   d. analytic coding.

7. Musical overtones are

   a. a series of complex waveforms.
   b. repetitions of the fundamental frequency at a constant intensity.
   c. multiples of the fundamental frequency.
   d. repetitions of the fundamental frequency at varying intensities.

8. The source of continuous low-pitched sounds is detected through *phase differences*, which is

   a. the time interval between the arrival at each ear of different portions of the oscillating sound wave.
   b. the time interval between the arrival at each ear of the same portion of the oscillating sound wave.
   c. the simultaneous arrival at each ear of the same portion of an oscillating sound wave.
   d. a or c, depending on the frequency of the stimulus.

9. Bilateral lesions of the auditory cortex of monkeys results in

   a. impaired ability to detect high frequency sounds.
   b. inability to detect intermittent, but not continuous sounds.
   c. gradual deafness.
   d. impaired ability to determine the source of a sound.

10. The semicircular canals

    a. respond to gravity.
    b. are part of the system of ossicles.
    c. are located in the sagittal, transverse, and horizontal planes in the head.
    d. respond to steady rotation of the head.

11. The vestibulo-ocular reflex depends upon vestibular connection to the

    a. third, fourth, and sixth cranial nerve nuclei.
    b. cerebellum.
    c. cortex.
    d. lower brain stem.

12. The gelatinous mass within the vestibular sacs shifts in response to movement of the

    a. cilia.
    b. otoconia.
    c. fluid in the cupula.
    d. fluid in the semicircular canal.

Lesson II: The Somatosenses, Gustation, and Olfaction

Read the interim summary on page 212-213 of your text to re-acquaint yourself with the material in this section.

Learning Objective 7-4   Describe the cutaneous receptors and their response to touch, temperature, and pain.

Read pages 205-209 and answer the following questions.

1. Identify the kind of information each of the following senses provides.

   a. cutaneous senses

   b. kinesthesia

    c. organic senses

2. List five stimuli to which the cutaneous senses respond.

    1.   4.

    2.   5.

    3.

3. a. List two functions of the skin.

    1.   2.

    b. List the three layers of the skin.

    1.   3.

    2.

4. Study Figure 7.23 in your text and then summarize information on cutaneous receptors by completing Table 1.

Name	Type of skin	Location	Type of information
1.	hairy		
2.	hairy		
3.	glabrous		
4.	glabrous		
5.	glabrous		

Table 1

5. a. Explain why the Pacinian corpuscle responds to vibration and not steady pressure. (Loewenstein and Mendelson, 1965)

    b. How does the bending of the tip of a cutaneous nerve ending produce a receptor potential? (Study Figure 7.24 in your text.)

6. Describe the phenomenon of adaptation, citing research by Nafe and Wagoner (1941).

7. Describe the cooperative roles played by the muscles and cutaneous receptors in detecting the physical characteristics of objects that a person touches.

8. a. Explain what we means when we say that feelings of warmth and coolness are relative.

    b. State two reasons why thermal receptors are difficult to study.

    c. Identify the receptors that appear to respond to temperature change. (Sinclair, 1981)

9. How does pain appear to be produced? (Besson et al., 1982)

Learning Objective 7-5   Describe the somatosensory pathways and the perception of pain.

Read pages 209-212 and answer the following questions.

1. Trace the somatosensory pathways with the help of Figure 7.25 in your text.

    a. In general, how do most of the somatosensory axons from the skin, muscles, and internal organs enter the central nervous system?

    b. Where do somatosensory axons from the head and face enter the CNS?

    c. Axons carrying precisely localized information ascend the dorsal columns in the white matter of the spinal cord to which brain region?

    d. These axons then cross the brain and ascend to which region? Where do axons from the thalamus project?

    e. Axons carrying poorly localized information follow a different route. What pathway do these axons follow to ascend to the ventral posterior nuclei of the thalamus?

2. Describe the arrangement of cells within the somatosensory cortex. (Mountcastle, 1957).

3. What did Dykes' (1983) review of research conclude about the primary and secondary somatosensory cortical area?

4. a. Explain why the perception of pain is beneficial to us.

    b. Give an example of an environmental event that can alter a person's perception of pain. (Beecher, 1959)

5. a. Describe some of the sensations reported by people who have had a limb amputated–the phantom limb phenomenon. (Melzak, 1992)

   b. Outline the classic explanation and treatment for this phenomenon.

   c. According to Melzak, what may account for phantom limb sensations? Cite the experiences of people with brain lesions or birth defects that support his suggestion.

6. a. List the two brain regions that effectively produce analgesia when electrically stimulated.

   1.   2.

   b. How has electrical stimulation been used to help humans suffering from chronic pain? (Kumar et al., 1990)

7. Study Figure 7.26 in your text and describe the neural circuit that mediates opiate-induced analgesia proposed by Basbaum and Fields (1978, 1984).

8. Finally, summarize research on reduced sensitivity to pain.

   a. classically conditioned analgesia (Maier et al., 1982)

   b. acupuncture (Mann et al., 1973; Gaw, 1975)

   c. behaviors important to survival (Komisaruk and Steinman, 1987; Whipple and Komisaruk, 1988)

   d. placebo (Levine et al., 1979)

Read the interim summary on page 218-219 of your text to re-acquaint yourself with the material in this section.

Learning Objective 7-6   Describe the four taste qualities, the anatomy of the taste buds and how they detect taste, and the gustatory pathway and neural coding of taste.

Read pages 213-218 and answer the following questions.

1. List the four qualities of taste.

   1.   2.

   3.   4.

2. How does flavor differ from taste?

3. Explain the biological significance of each of the four taste qualities.

4. Study Figure 7.27 in your text and describe the
    a. location and appearance of the papillae.

    b. location of the taste buds in relation to the papillae.

    c. function of the trenches.

5. Describe the regions of taste sensitivity on the tongue. (See Figure 7.28 in your text.)

6. Describe the location, function, life-span, and replacement of the taste receptors. (Beidler, 1970)

7. In general, how are different taste sensations produced by a molecule of a tasted substance?

8. List the best stimulus for the taste receptors for
    a. saltiness.   c. bitterness.

    b. sourness.   d. sweetness

9. Carefully study Figure 7.29 in your text and explain the changes that may occur when a tasted molecule binds with a taste receptor.
    a. salty

    b. sour

    c. bitter

d. sweet

10. What other two taste qualities have been proposed by researchers?

11. List the three cranial nerves that carry gustatory information to the brain and indicate which regions they serve.

1.   3.

2.

12. Begin with taste receptors on the tongue and continue to trace the gustatory pathway.

a. Name the first station on this pathway.

b. Where do the axons go from there?

c. What other nerve sends information to this nucleus? (Beckstead et al., 1980)

d. Where do thalamic taste-sensitive neurons project? (Pritchard et al., 1986)

e. Finally, what other brain regions receive gustatory information? (Nauta, 1964; Russchen et al., 1986)

13. Summarize the results of research on the neural coding of taste in the

a. chorda tympani (Nowlis and Frank, 1977).

b. gustatory cortex (for example, Scott et al., 1991; Rolls, 1995).

Read the interim summary on pages 222-223 of your text to re-acquaint yourself with the material in this section.

Learning Objective 7-7   Describe the major structures of the olfactory system, explain how odors may be detected, and describe the patterns of neural activity produced by these stimuli

Read pages 219-222 and answer the following questions.

1. List the two functions of most odorous substances.

1.   2.

2. a. Where are olfactory receptors found?

b. Where is the olfactory epithelium found? (See Figure 7.32 in your text.)

b. Where is the olfactory epithelium found? (See Figure 7.32 in your text.)

3. Continue to refer to Figure 7.32 and explain what happens when you hold a rose to your nose and sniff it.

   a. What sweeps the air through the nasal cavity toward the olfactory receptors?

   b. Where are the cell bodies of olfactory receptors located?

   c. Where does a bipolar olfactory cell process project?

   d. What may be the function of free nerve endings of the trigeminal nerve found in the olfactory mucosa?

   e. Where do the olfactory receptor cells send axons?

   f. Where to the receptor cell axons synapse, and with the dendrites of what cells?

   g. Where olfactory tract axons project?

   h. With regard to olfaction, what may be the function of the

      1. orbitofrontal cortex.

      2. hypothalamus.

4. a. What is the role of the G_{olf} protein? (For example, Jones and Reed, 1989; Nakamura and Gold, 1987)

   b. What did Buck and Axel (1991) discover?

5. Why it is unlikely that each odor substance is detected by it own receptor?

6. Study Figure 7.33 and review the relationship between the receptors, olfactory neurons, and the glomeruli.

   a. The cilia of each olfactory neuron contain _____ type of receptor. Each glomerulus receives information from approximately _____ _____ different olfactory receptor cells.

   b. What did Ressler et al. (1994) discover about the relationship between receptor cells and glomeruli?

7. How can we use a relatively small number of receptors to detect so many different odorants? (Refer to Figure 7.34 in your text.)

Lesson II Self Test

1. We experience feelings of pressure through

    a. the cutaneous senses.
    b. kinesthesia.
    c. the organic senses.
    d. the vestibular senses.

2. The _____, which are found in _____ skin, are the largest sensory end organs in the body.

    a. Pacinian corpuscles; hairy
    b. Ruffini corpuscles; hairy
    c. Pacinian corpuscles; glabrous
    d. Ruffini corpuscles; glabrous

3. The reason why people ignore the pressure from a ring or a belt that they wear daily is that receptor cells

    a. become fatigued from constant information.
    b. adapt to constant stimulation.
    c. degenerate from constant pressure and are not replaced.
    d. are constricted by the constant pressure.

4. Somatosensory cortex

    a. is divided into maps that allocate the most tissue to the regions with the greatest surface area such as the skin of the back.
    b. is represented by five horizontal layers.
    c. is arranged in columns that respond to a particular type of stimulus.
    d. has yet to be represented on functional maps.

5. The most effective locations for producing analgesia using electrical brain stimulation are the

    a. nucleus of the solitary tract and the nucleus raphe magnus.
    b. trigeminal nerve and the rostroventral medulla.
    c. periaqueductal gray matter and the rostroventral medulla.
    d. nucleus raphe magnus and the posterior nuclei of the thalamus.

6. Electrical brain stimulation apparently produces analgesia by stimulating the release of

    a. prostaglandins.
    b. morphine.
    c. histamines.
    d. endogenous opiates.

7. Substances that taste sweet are first tasted _____ of the tongue.

    a. on the tip
    b. in the middle
    c. at the back
    d. along the sides

8. Gustatory information from the cranial nerves is sent directly to the

    a. parabrachial nucleus of the pons.
    b. thalamic taste area.
    c. hypothalamus.
    d. nucleus of the solitary tract.

9. Studies have confirmed that _____ and _____ substances activate a G-protein and second messenger.

    a. sweet, bitter
    b. sweet, umami
    c. sour, salty
    d. bitter, salty.

10. Olfactory receptors are located in the

    a. olfactory bulbs.
    b. cribriform plate.
    c. olfactory epithelium.
    d. olfactory tracts.

11. Electrical recordings of neural responses to olfactory stimuli indicate that

    a. response patterns to odors remain constant.
    b. half the sampled neurons in one region responded to only one particular odor.
    c. olfactory receptors exist for every detectable odor.
    d. the metabolic rate of olfactory receptors is not affected by odor

12. All of the above occur during transduction of olfactory information except:

a. A special G protein is activated in the olfactory cilia.

b. The membrane of the cell depolarizes.

c. Cyclic AMP opens potassium channels.

d. Sodium channels open.

Answers for Self Tests

Lesson I

1. a Obj. 7-1
2. c Obj. 7-1
3. d Obj. 7-1
4. b Obj. 7-1
5. a Obj. 7-1
6. c Obj. 7-2
7. b Obj. 7-2
8. d Obj. 7-2
9. d Obj. 7-2
10. c Obj. 7-3
11. b Obj. 7-3
12. c Obj. 7-3

Lesson II

1. a Obj. 7-4
2. c Obj. 7-4
3. b Obj. 7-4
4. c Obj. 7-5
5. c Obj. 7-5
6. d Obj. 7-5
7. a Obj. 7-6
8. d Obj. 7-6
9. a Obj. 7-6
10. c Obj. 7-7
11. b Obj. 7-7
12. c Obj. 7-7

CHAPTER 8
Control of Movement

Lesson I: Muscles and the Reflexive Control of Movement

Read the interim summary on pages 229-230 of your text to re-acquaint yourself with the material in this section.

Learning Objective 8-1   Describe the three types of muscles found in the bodies of mammals, and explain the physical basis of muscular contraction.

Read pages 225-229 and answer the following questions.

1. List the three types of muscles found in the bodies of mammals.

   1.   2.   3.

2. How are skeletal muscles attached to the bones?

3. a. Describe flexion and give an example.

   b. Now describe extension and give an example.

4. Add labels to Figure 1 on the next page. (See Figure 8.1 in your text.)

5. Identify the functions of the following structures:

   a. extrafusal muscle fiber.

   b. alpha motor neuron.

   c. sensory endings in the central region of the intrafusal muscle fiber.

   d. gamma motor neuron.

6. What is the relationship between precision of movement and the number of muscle fibers served by a single axon of the alpha motor neuron?

7. List the three components of a motor unit.

   1.   2.   3.

8. a.   List the two kinds of protein filaments found in a myofibril and describe their function. (Return to Figure 8.1. See the marginal definitions.)

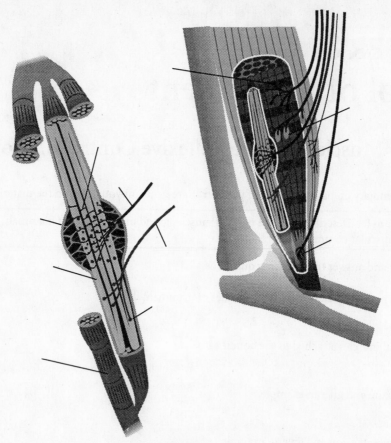

Figure 1

9. A(n) _____ _____ is the synapse between the terminal button of a(n) _____ _____ and the membrane of a(n) _____ _____. The terminal buttons of the _____ synapse on _____ _____.

10. Identify the following events involved in a muscular contraction.

    a. the neurotransmitter that is released by the terminal buttons when an axon fires

    b. the change in the postsynaptic membrane that is produced by the transmitter substance

    c. the unalterable effect on the membrane of the muscle fiber

    d. the effect on calcium channels and ionic movement

11. What is the role of actin and myosin in this process? (Study Figure 8.2 in your text.)

12. Cite two reasons why the physical effects of a muscular contraction last longer than the action potential. (See Figure 8.3 in your text.)

13. What determines the strength of a muscular contraction?

14. a. Explain the distinction between the functions of the stretch receptors of the intrafusal muscle fibers and those of the Golgi tendon organ.

    b. To be sure that you understand this distinction, describe the response of the intrafusal muscle fibers and the Golgi tendon organ to various kinds of movement which are illustrated in Figure 8.4 in your text.

15. There are two types of smooth muscle. List some of the places where each is found and note what initiates contractions.

    1. multiunit smooth muscle

    2. single-unit smooth muscle

16. a. What factors modulate heart rate?

    b. What initiates the heartbeat?

Read the interim summary on pages 233-234 of your text to re-acquaint yourself with the material in this section.

Learning Objective 8-2   Explain the monosynaptic stretch reflex, the gamma motor system, and the contribution of the Golgi tendon organ.

Read pages 230-233 and answer the following questions.

1. a. Describe the patellar reflex.

   b. How do we know this reflex does not involve the brain?

2. Study Figure 8.5 in your text and trace the pathway of the monosynaptic stretch reflex.

   a. Where is the only synapse in this reflex circuit?

   b. When a weight is placed in the hand, what happens to the activity of the muscle spindles?

   c. Briefly explain how the monosynaptic stretch reflex aides in the control of posture. (See Figure 8.6 in your text.)

3. Explain the relationship between activity of the gamma motor system and the degree of sensitivity to stretch by the intrafusal muscle fibers.

4. When the brain initiates limb movement

    a. what two sets of motor neurons in the spinal cord are stimulated?

    b. and there is little resistance, how will the extrafusal and the intrafusal muscle fibers respond? the afferent axons of the muscle spindle?

    c. and unexpected resistance is encountered, how will extrafusal and intrafusal muscle fibers now respond?

5. a. What do the more sensitive afferent axons from the Golgi tendon organ detect? the less sensitive ones?

    b. Now study Figure 8.7 in your text and trace this polysynaptic reflex circuit originating at the Golgi tendon organ. Where do the terminal buttons of these less sensitive axons synapse and what is their function?

    c. Where do the terminal buttons of the interneurons form synapses?

    d. What do they secrete and what is the effect of its release?

    e. What is the function of this reflex pathway?

6. What may be the consequence of blocking the Golgi tendon organ with a local anesthetic?

7. a. Describe the cause and characteristics of decerebrate rigidity in a cat.

    b. Outline the neural mechanism which affects the stretch reflex. Be sure to mention the clasp-knife reflex in your answer.

8. a. A stretch reflex excites the _____ and inhibits the _____.

    b. Now study Figure 8.8 in your text and explain why this occurs.

Read the interim summary on pages 250-251 of your text to re-acquaint yourself with the material in this section.

Learning Objective 8-3   Describe the organization of motor cortex, and describe the four principal motor tracts and the movements they control.

Read pages 234-240 and answer the following questions.

1. a. What do we mean when we say that the primary motor cortex shows somatotopic organization?

    b. Study Figure 8.9 in your text and explain why some features of the motor homunculus are exaggerated in size.

2. Examine the relationship between the primary motor cortex and the frontal association cortex.

    a. What is the principal cortical input to the primary motor cortex?

    b. Identify and describe the location of two other regions that also send axons to the primary motor cortex. Compare their inputs and outputs.

3. a. Where are most complex behaviors planned? executed?

    b. Study Figure 8.10 in your text and summarize the kinds and sources of sensory information the posterior association cortex and the frontal association cortex receive.

    c. How does this sensory information contribute to the control of movement?

4. What is the anatomical and functional relationship between the primary motor cortex and the primary somatosensory cortex?

5. a. What tasks did Evarts (1974) teach subject monkeys? (Study Figure 8.11 in your text.)

    b. What did single cell recordings indicate about the

    1. relationship between movement and the rate of firing by the cell?

    2. the role of the postcentral gyrus in hand and finger movements?

6. Neurons in the _____ _____ _____ control movement through two groups of descending tracts located in the _____ _____ of the spinal cord. The _____ group consists of the _____ tract, the _____ tract, and the _____ tract and is principally involved in the control of _____ _____ movements. The _____ group consists of the _____ tract, the _____

tract, the _____ tract, and the _____ _____ tract and controls more _____ movements.

7. Let's trace these pathways, shown in Figures 8.12 and 8.13 in your text, through which the primary motor cortex controls movement beginning with the lateral group of descending pathways.

   a. Where do axons in the corticospinal tract originate? terminate?

   b. What other brain regions send axons through this pathway?

   c. Describe the pathway that axons follow to the cerebral peduncles?

   d. At what point are the axons of the corticospinal called the pyramidal tracts.

   e. Most fibers now cross to the other side of the brain to form what tract?

   f. The remaining fibers descend to form what other tract?

   g. Where do the axons of the lateral corticospinal tract and the ventral corticospinal tract originate? terminate? (Study the light and dark blue lines in Figure 8.12.)

   h. What kind of movement does each tract control?

8. Describe the recovery of movement and any difficulties Lawrence and Kuypers (1968a) observed in monkeys whose pyramidal tracts (corticospinal tracts) were cut.

   a. 6-10 hours after recovery from anesthesia

   b. the day after surgery

   c. six weeks later

9. What do their results indicate about the organization of the control of fingers, posture, and locomotion? the same behavior in different circumstances?

10. a. Where does the corticobulbar pathway terminate? (Study the green lines in Figure 8.12.)

    b. What functions does it control?

11. a. Where does the rubrospinal tract originate and synapse? (Study the red lines in Figure 8.12.)

    b. Which muscle groups are controlled by this pathway? are not controlled?

12. In another study Lawrence and Kuypers (1968b) destroyed the rubrospinal tract unilaterally in some of the monkeys who had previously received bilateral pyramidal tract lesions.

    a. Describe the eating behavior of these monkeys.

    b. What do the results suggest about the function of the rubrospinal pathway?

13. List again the four tracts that make up the ventromedial group of descending pathways. Next to each tract indicate where its cell bodies are located and its function.

    1.

    2.

    3.

    4.

14. Lawrence and Kuypers (1968b) also cut the ventromedial fibers of some of the monkeys who had previously received bilateral pyramidal tract lesions.

    a. Describe the posture, locomotion, and eating movements of these monkeys.

    b. What do the results suggest about the function of ventromedial pathways?

To review these pathways, their locations, and the muscle groups they control, study Table 8.1 in your text.

Lesson I Self Test

1. The number of muscle fibers that serve a single axon depends on

    a. the size of the body part that must be moved.
    b. the precision with which a muscle can be controlled.
    c. whether its controls flexion or extension.
    d. its sensitivity to stretch.

2. The three components of a motor unit are

    a. an alpha motor neuron; its axon; and associated extrafusal muscle fibers.

    b. a gamma motor neuron; its axon; and associated intrafusal muscle fibers.
    c. an alpha motor neuron; its terminal buttons; and associated neurotransmitter.
    d. muscles; tendons; and associated stretch receptors.

3. During a depolarization of the muscle fiber, which event does not occur?

    a. Calcium enters the cytoplasm.
    b. The movement of actin causes muscle fiber to shorten.

c. The myosin crossbridges move, shortening the muscle fiber.

d. Myofibrils extract energy provided by the mitochondria.

4. The Golgi tendon organ detects

a. muscle length.
b. strength of muscular contraction.
c. the total amount of stretch.
d. rate of muscular contraction.

5. The monosynaptic stretch reflex

a. initiates limb withdrawal in response to pain.
b. helps compensate for changes in weight that cause limb movement.
c. maintains muscles in a constant state of contraction.
d. is the simplest neural pathway and has little utility.

6. A polysynaptic reflex

a. can be demonstrated by tapping the patellar tendon.
b. contain no interneurons between the sensory neurons and the motor neuron.
c. simplifies the role of the brain in controlling movement.
d. limits the amount of muscular contraction to prevent injury.

7. Muscle spindles are sensitive to changes in

a. blood levels of calcium.
b. tendon stretch.
c. muscle tension.
d. muscle length.

8. When a stretch reflex is elicited in the agonist muscle it _____quickly causing the antagonist to _____.

a. contracts; lengthen
b. lengthen; contract
c. releases; release at the same rate

d. releases; pull back

9. Stimulation studies of primary motor cortex indicate that the largest amount of cortical area is devoted to movements of

a. arms and legs.
b. head and neck.
c. fingers and speech muscles.
d. trunk and genitalia.

10. The principal cortical input to the primary motor cortex is

a. temporal cortex.
b. prefrontal cortex.
c. frontal association cortex.
d. posterior association cortex.

11. The fact that monkeys with bilateral pyramidal tract lesions had no trouble opening their hands when climbing but had difficulty opening their hands when eating indicates that

a. these monkeys can still successfully forage for food.
b. damaged neural pathways that mediate large movements can regenerate, but neural pathways that mediate precise movements cannot.
c. the ability to use the hands to flee (climb) conveys a greater selective advantage to the species than feeding.
d. the same behavior can be controlled by different brain mechanisms in different contexts.

12. The reticulospinal tracts controls

a. movements of the fingers and hands.
b. several autonomic functions such as respiration.
c. movements of forelimb and hindlimb muscles.
d. posture and righting reflexes.

Lesson II: The Apraxias, the Basal Ganglia, the Cerebellum, and the Reticular Formation

Read the interim summary on pages 250-251 in your text to re-acquaint yourself with the material in this section.

Learning Objective 8-4   Describe the symptoms and causes of limb apraxia and constructional apraxia.

Read pages 240-242 and answer the following questions.

1. Define *apraxia* in your own words and explain the difference between apraxia and paralysis.

2. List the four kinds of apraxia and the associated deficit.

    1.   3.

    2.   4.

3. a. More specifically, list the three kinds of movement difficulties patients with limb apraxia may exhibit.

    1.   3.

    2.

   b. Describe how a patient with limb apraxia is tested orally, beginning with the easiest task.

   c. If the patient cannot understand speech, how is the apraxia evaluated? (Heilman et al., 1983)

4. Summarize the cause and physical disabilities for each of the following types of limb apraxia.

   a. Callosal apraxia

       1. Lesion site (Study Figure 8.14, lesion A in your text.)

       2. Affected neural circuits

       3. Affected limb

   b. Sympathetic apraxia

       1. Lesion site (Study Figure 8.14, lesion B)

       2. Affected limbs

   c. Left parietal apraxia

       1. Lesion site (Study Figure 8.14, lesion C)

       2. Affected limbs

5. a. According to Mountcastle et al. (1975), what may be the primary motor function of the region of the parietal lobes?

   b. What is the presumed function of the right parietal lobe? left parietal lobe?

6. Study Figure 8.15 in your text and then explain how the command to reach for an object is processed by the brain. Begin your explanation right after the person hears the request to do something.

7. a. What kind of lesion results in constructional apraxia?

b. What are some tasks that people with constructional apraxia can and cannot perform successfully? (See Figure 8.16 in your text.)

Learning Objective 8-5   Discuss the anatomy and function of the basal ganglia, and its role in Parkinson's disease and Huntington's chorea.

Read pages 242-247 and answer the following questions.

1. a. List the motor nuclei of the basal ganglia. (Study Figure 8.17a in your text.)

    1.   2.   3.

    b. Now list some of the nuclei associated with the basal ganglia.

    1.   2.   3.

    c. Finally, list the afferent and efferent connections of the basal ganglia.

    1. afferent connections

    2. efferent connections

    c. Which motor systems do the basal ganglia thus influence?

    1.   2.

2. Trace the connections between the basal ganglia and the cortex shown in Figure 8.17 b.

    a. Where do primary motor cortex and primary somatosensory cortex send axons?

    b. Where does the putamen send axons?

    c. And where does the globus pallidus send axons?

3. What do we mean when we say the information in this circuit is represented somatotopically?

4. What other important structures send information to the basal ganglia?

5. Let's look more closely at the nature of the connections in the cortical-basal ganglia loop that you have just traced.

    a. Which neurotransmitter is secreted by the excitatory neurons in this loop? the inhibitory neurons?

    b. What kind of axons does the putamen receive from the cerebral cortex?

    c. Where does the putamen, in turn, send axons and what kind of axons are they?

    d. Where does the GP_i send axons and what kind of axons are they?

e. And where does the VA/VL thalamus send axons and what kind are these?

f. Carefully explain why the net effect of this loop is excitatory.

g. Finally, where in this loop does the GP$_e$ send axons? the globus pallidus?

6. Briefly describe the primary symptoms of Parkinson's disease and how they disrupt such activities as getting up, walking, writing, and maintaining balance.

7. Explain how we know that tremor and rigidity are not the cause of slowness of motion.

8. Study Figure 8.17a in your text and then add labels to the ovals and the leader lines in Figure 2 on the next page. Add the missing arrows, indicating excitatory connections with black lines and inhibitory connections with red or broken lines.

9. a. Why does treatment of Parkinson's disease with L-DOPA eventually become less effective?

b. When drug therapy fails, what other treatments are possible?

c. What do neurosurgeons hope to achieve by transplanting fetal tissue into the brains of people with Parkinson's disease? How successful is this surgery?

10. a. Explain why destruction of the GP$_i$ during a pallidotomy might relieve the symptoms of Parkinson's disease.

b. What was a side effect of early operations?

c. What developments caused this surgical procedure to be abandoned and then resumed?

d. How have technological advances improved the outcome of pallidotomies? (Grafton et al., 1995)

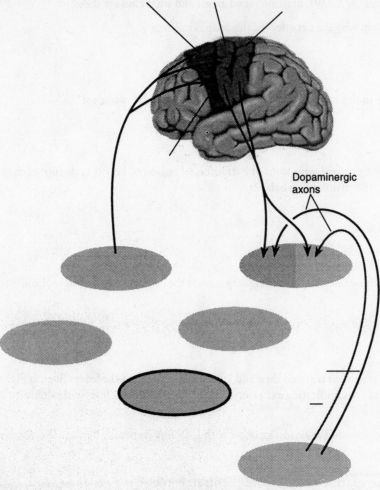

Dopaminergic axons

Figure 2

11. Parkinson's disease may be caused by toxins. Identify three possible sources of toxins and cite evidence to support your answer. (Langston et al., 1983; Langston et al., 1984)

12. Carefully explain the effects of deprenyl to explain its promise as a treatment for Parkinson's disease. (Tetrud and Langston, 1989)

13. a. What is the cause of Huntington's chorea. (See Figures 8.17b and 8.18 in your text.)

    b. Where does neural degeneration appear to begin and how does it progress?

    c. At what age do the symptoms usually begin?

    d. How does progressive degeneration affect movement?

e. Briefly describe the location of the gene that causes Huntington's chorea, its particular defect, and how this defect may lead to the death of these neurons in the brain. Be sure to mention huntingtin in your answer.(Collaborative Research Group, 1993)

f. How may the abnormal huntingtin interact with other proteins? (Burke et al., 1996; Li et al., 1995; 1996; Dawson et al., 1993)

Learning Objective 8-6   Discuss the role of the cerebellum and the reticular formation in the control of movement.

Read pages 247-250 and answer the following questions.

1. In general, what kind of deficits occur if the cerebellum is damaged?

2. Describe neural circuits found in these regions of the cerebellum. Identify the input(s) and output(s) to each region, and the types of movements that are controlled. (Study Figures 8.19 and 8.20 in your text.)

   a. flocculonodular lobe

   b. vermis (Be sure to mention the fastigial nucleus.)

   c. intermediate zone (Be sure to mention the interposed nuclei.)

   d. lateral zone (Be sure to mention the pontine nucleus.)

3. When motor cortex initiates a movement, how does the cerebellum respond? (Study Figure 8.20)

4. List some of the movement deficit resulting from damage to the

   a. flocculonodular lobe or vermis.

   b. intermediate zone.

   c. lateral zone (Be sure to describe decomposition of movement and ballistic movements.)

5. According to Kornhuber (1974) what may be a primary function of the cerebellum?

6. Describe research findings (Thach, 1978) that support earlier clinical observations (Holmes, 1939) that the cerebellum appears to integrate successive sequences of movements.

7. Where is the reticular formation located?

8. List some of the functions of the reticular formation.

    1.   3.

    2.

9. Describe the results of the following studies of the reticular formation.

    a. stimulation of the mesencephalic locomotor region (Shik and Orlovsky, 1976)

    b. recordings from single neurons in freely moving cats (Siegel and McGinty, 1977)

Lesson II Self Test

1. Apraxia is the inability to

    a. properly execute a learned skilled movement.
    b. benefit from practicing a skilled movement.
    c. perceive sequences of skilled movements.
    d. resume performing a skilled movement that has been interrupted.

2. Limb apraxia is assessed by asking a patient to

    a. perform a movement.
    b. describe how he or she would move in a particular situation.
    c. teach someone else a movement.
    d. list components of a motion in sequential order.

3. Patients with callosal apraxia are able to perform a requested movement with their right arm, but not their left because

    a. the lesion has paralyzed the left arm.
    b. most people are right handed and the lesion does not affect "handedness."
    c. the anterior corpus callosum has been damaged and the right and left premotor areas can no longer communicate.
    d. Wernicke's area has been damaged and requests to perform a movement are only partially understood.

4. Patients with constructional apraxia have difficulty

    a. pantomiming particular actions.

    b. "constructing" the proper sequence of action when shown a series of pictures, in random order, of the components of a motion.
    c. controlling the movements of their hands and arms.
    d. building shapes using toy building blocks.

5. The primary deficit in constructional apraxia appears to involve

    a. motor impairment of the hands and arms.
    b. the ability to perceive and imagine geometric shapes.
    c. difficulty in sequencing of actions.
    d. the inability to follow instructions.

6. The components of the basal ganglia are the

    a. ventrolateral nucleus; the ventral anterior nucleus; and the pontine nucleus.
    b. premotor cortex; the primary motor cortex; and the supplementary motor area.
    c. the globus pallidus; the substantia nigra; and the subthalamic nuclei.
    d. caudate nucleus; the putamen; and the globus pallidus.

7. Parkinson's disease is characterized by _____ movements and Huntington's chorea is characterized by _____ movements.

    a. slow; uncontrollable
    b. smooth; slow
    c. rigid; smooth
    d. slow; smooth

8. A pallidotomy destroys the

   a. substantia nigra.
   b. internal division of the globus pallidus.
   c. external division of the globus pallidus.
   d. putamen.

9. Huntington's chorea is caused by a defective gene that

   a. causes the faulty conversion of MPTP into toxic MPP.
   b. results in abnormally high levels of GADPH.
   c. halts nitric oxide synthesis.
   d. produces a protein with an elongated stretch of glutamine.

10. The cerebellum consists of two _____ with the _____ located on the midline.

    a. lobes; intermediate zone
    b. zones; corpus callosum

c. sets of nuclei; interposed nuclei
d. hemispheres; vermis

11. If a patient complains to a physician that he or she has recently been having difficulty maintaining balance, the physician may suspect a lesion in the

    a. basal ganglia.
    b. lateral zone of the cerebellum.
    c. red nucleus.
    d. flocculonodular lobe of the cerebellum.

12. Stimulation of the mesencephalic locomotor region of the reticular formation causes cats to

    a. sit down.
    b. tremble.
    c. pace.
    d. stare.

Answers for Self Tests

Lesson I		
1.	b	Obj. 8-1
2.	a	Obj. 8-1
3.	b	Obj. 8-1
4.	b	Obj. 8-1
5.	b	Obj. 8-2
6.	d	Obj. 8-2
7.	d	Obj. 8-2
8.	a	Obj. 8-2
9.	c	Obj. 8-3
10.	c	Obj. 8-3
11.	d	Obj. 8-3
12.	c	Obj. 8-3

Lesson II		
1.	a	Obj. 8-4
2.	a	Obj. 8-4
3.	c	Obj. 8-4
4.	d	Obj. 8-4
5.	b	Obj. 8-4
6.	d	Obj. 8-5
7.	a	Obj. 8-5
8.	b	Obj. 8-5
9.	d	Obj. 8-5
10.	d	Obj. 8-6
11.	d	Obj. 8-6
12.	c	Obj. 8-6

CHAPTER 9
Sleep and Biological Rhythms

Lesson I: Sleep and its Functions

Read the interim summary on page 258 of your text to re-acquaint yourself with the material in this section.

| Learning Objective 9-1   Describe the course of a night's sleep: its stages and their characteristics. |

Read pages 253-258 and answer the following questions.

1.   Explain why sleep is properly considered to be a behavior.

2.   Briefly describe the appearance of a sleep laboratory.

3.   List the physiological functions that are monitored during a night of sleep in a sleep laboratory and explain how they are measured.

4.   Name and describe the two activity patterns that characterize the waking EEG of a normal person and identify the behavioral state that each accompanies.

   1.   2.

5.   Study Figure 9.2 in your text and fill in the table below.

Stage	Name of EEG pattern	Description of EEG pattern
Resting		
Alert		
Stage 1		
Stage 2		
Stage 3		
Stage 4		
REM sleep		

6. Describe the characteristics of neural synchrony and desynchrony.

7. Discuss the EEG record, the frequency, and possible significance of sleep spindles and K complexes. (Bowersox et al., 1985; Niiyama et al., 1995,1996)

8. During which stage of sleep will a sleeping person who is awakened insist that he or she had not yet fallen asleep?

9. Which stage of sleep contains the most delta activity?

10. Summarize the changes that occur during REM sleep in the

    a. EEG   c. EMG

    b. EOG

11. a. What does the acronym REM stand for?

    b. Which stages of sleep are called non-REM sleep? slow-wave sleep?

    c. What stimuli will cause a sleeper to awaken from stage 4 sleep? REM sleep?

12. Compare the dreams of sleepers during REM and slow-wave sleep.

13. Study Figure 9.3 in your text and describe the typical pattern of sleep stages that occurs during a night's sleep.

    a. Normal sleep alternates between what two kinds of sleep?

    b. Approximately how long is each cycle?

    c. How many periods of REM sleep occur each night?

    d. When does most slow-wave sleep occur?

    e. Which stages of sleep predominate as morning approaches?

14. a. What name did Kleitman give to the mechanism that regulates both the alternating pattern of REM and slow-wave sleep and the activity cycle during waking? (Kleitman 1982)

    b. What observation of infant behavior first suggested the existence of this 90-minute activity cycle? (Kleitman, 1961)

15. Explain how a response of male subjects that occurs during REM sleep has been used to assess the causes of impotence. (Karacan et al., 1978)

16. Review by listing the principal characteristics of slow-wave and REM sleep that you have learned so far, and then compare your list with Table 9.1 in your text.

REM Sleep	Slow-Wave Sleep

17. Refute these incorrect assertions.

   a. Sleep is a state of unconsciousness.

   b. Some people never dream.

18. Describe the cerebral blood flow during REM sleep (Madsen et al., 1991). What does this suggest about REM dreams? (Melges, 1982; Hobson, 1988)

19. How may the eye movements characteristic of REM sleep be related to dreaming? (Roffwarg et al., 1962) Cite research to support your answer. (Miyauchi et al., 1990)

20. What other brain mechanisms may be active during a dream?

Read the interim summary on pages 265-266 of your text to re-acquaint yourself with the material in this section.

Learning Objective 9-2   Review the hypothesis that sleep is an adaptive response.

Read pages 258-259 and answer the following questions.

1. Sleep is a _____ phenomenon among _____. However, only warm-blooded vertebrates experience _____ sleep.

2. According to Webb (1975, 1982) what may be the function of sleep? What evidence supports his hypothesis?

3. Summarize evidence that sleep is more than an adaptive response and may be physiologically necessary.

   a. Describe the sleep patterns of the Indus dolphin. (Pilleri, 1979)

b. Study Figure 9.5 in your text and describe the sleep patterns of the cerebral hemispheres of the bottlenose dolphin. (Mukhametov, 1984)

c. Now explain why these unusual sleep patterns suggest sleep is more than an adaptive response?

Learning Objective 9-3   Review the hypothesis that sleep serves as a period of restoration by discussing the effects of sleep deprivation, exercise, and mental activity.

Read pages 259-262 and answer the following questions.

1. Explain the rationale for using sleep deprivation to study the functions of sleep.

2. What did a review of over fifty sleep studies indicate about the effects of sleep deprivation on stress? on the ability to perform physical exercise? (Horne, 1978)

3. Summarize the experience of a teenager who obtained a place in the *Guinness Book of World Records* by comparing

   a. the total number of hours of his enforced wakefulness with the number of hours he slept at the end of his record attempt.

   b. the percentage of recovery of stage 1 and 2 sleep and slow-wave (stage 4) and REM sleep. (Gulevich et al., 1966)

   c. the importance of each sleep stage.

4. Describe further research on slow-wave sleep. (Sakai et al., 1979; Buchsbaum et al., 1989; Bonnet and Arand, 1996).

5. What is the progression of the disease *fatal familial insomnia*, and what does it tell us about the functions of sleep?

6. a. See Figure 9.6 in your text and describe the apparatus and procedure that Rechtschaffen et al. (1983, 1989) developed to keep rats awake and exercising. Be sure to use the term *yoked control* in your answer.

   b. How successful were the experimenters in producing sleep deprivation?

c. Summarize the effects of sleep deprivation on both the experimental (sleep-deprived) rats and the yoked-control rats.

d. What do the results suggest about the importance of sleep?

7. Explain the rationale for evaluating a possible restorative function of sleep by studying the affects of daytime activity on nighttime sleep.

8. Summarize changes in slow-wave sleep of

a. healthy subjects after six weeks of bed rest. (Ryback and Lewis, 1971)

b. completely immobile quadriplegics and paraplegics. (Adey et al., 1968)

9. When exercise does increase slow-wave sleep, what is an important variable? (Horne, 1981, 1988)

10. Describe how Horne and Moore (1985) studied the effect of body temperature during exercise on subsequent slow-wave sleep.

a. What were the two experimental procedures?

b. How did the experimental procedures affect the slow-wave sleep of subjects?

c. What does Horne now believe is the more important variable?

11. Explain the results of follow-up research that tends to support his conclusion. (Horne and Harley, 1989)

12. a. What is the effect of tasks that demand mental activity and alertness on the brain? In what part of the brain is the effect most significant?

b. Where is delta activity most intense during slow-wave sleep?

c. What does this suggest about the role of sleep?

d. Describe other research that supports this interpretation. (Kattler et al., 1994)

13. Describe how Horne and Minard (1985) studied the affects of increased mental activity on sleep.

a. What were subjects first told they would be doing during the experiment, and how were the plans changed?

b. What changes were observed in their slow-wave sleep that night, and what do the results suggest about the function of sleep?

14. Several researchers have suggested a link between thermoregulation and sleep.

   a. Summarize the hypothesis advanced by Berger and Phillips (1995) regarding the relationship between endothermy and sleep.

   b. Review the research by Horne and Reid (1985) and Morairty et al. (1988). How do they provide evidence of a link between thermoregulation and sleep?

   c. Horne (1992) finds fault with Berger and Phillips' hypothesis. Why?

15. Finally, what evidence suggests sleep may serve different purposes in different species?

Learning Objective 9-4   Discuss the functions of REM sleep.

Read pages 263-265 and answer the following questions.

1. Review the physiological changes that occur during REM sleep to

   a. the eyes.   c. breathing.

   b. the heart rate.   d. the brain.

2. When subjects are deprived of REM sleep, what do researchers observe

   a. as deprivation progresses? (Dement, 1960)

   b. a few days later, when subjects are permitted to sleep normally? Be sure to use the term *rebound phenomenon* in your answer.

3. Briefly state the hypothesis concerning the function of REM sleep that falls into each of these categories.

   a. vigilance (Snyder, 1966)

   b. learning (Greenberg and Pearlman, 1974; Crick and Mitchison, 1983, 1995)

   c. species-typical reprogramming (Jouvet, 1980)

   d. brain development (Roffwarg et al., 1966)

Let's review the evidence for each of these hypotheses.

4. a. How does REM sleep deprivation affect learning a new task, especially if it is complicated? (McGrath and Cohen, 1978; Smith, 1985)

   b. and how does learning a new task affect REM sleep?

5. Does any evidence exist that would support the vigilance and reprogramming hypotheses? Why or why not?

6. Discuss the characteristics of the REM sleep of the following, and the implications for the development hypothesis.

   a. guinea pigs, rats, cats, and humans.

   b. newborns infants, 6 month old babies, and older children.

7. a. What did Mirmiran (1995) and his colleagues observe in rats whose REM sleep had been suppressed during infancy?

   b. What did Marks et al. (1995) find in animals with brain lesions that disrupted one of the phenomena of REM sleep?

   c. Do these studies provide support for the brain development hypothesis? Why or why not?

8. If the function of REM sleep is to promote brain development, why might adults have REM sleep?

9. According to Smith's (1996) research, what is the relationship between memory formation and REM sleep deprivation? Describe specific evidence from his research.

10. Describe how Bloch et al. (1977) studied the effect of learning on the REM sleep of rats.

    a. What task were rat subjects taught to perform?

    b. Study Figure 9.7 in your text and describe how subsequent REM sleep was related to

       1. the learning that had occurred that day.

       2. daily performance.

11 a. How is human learning and remembering affected by REM sleep deprivation?

b. Compare the amount of REM sleep of retarded children and gifted children. (Dujardin et al., 1990)

c. How does the amount of REM sleep of college students change during exam week? (Smith and Lapp, 1991)

12 Study Figure 9.8 in your text and describe the two possible explanations for the relation between waking, slow-wave sleep and REM sleep.

13. Finally, what evidence suggests REM sleep may not be necessary for survival? (Lavie et al., 1984)

Lesson I Self Test

1. The waking EEG is characterized by

    a. occasional delta activity.
    b. periods of alpha and beta activity.
    c. regular changes in heart rate, blood pressure and respiration.
    d. bursts of K complexes.

2. Sleeping subjects who are awakened during REM sleep almost always

    a. report narrative dreams.
    b. insist they were still awake.
    c. report frightening situations.
    d. report they were not dreaming.

3. REM sleep

    a. almost always follows a period of slow-wave sleep.
    b. occurs four or five times during an 8-hour sleep and lasts approximately 90 minutes.
    c. contains more than 50 percent delta activity.
    d. is the deepest stage of sleep.

4. All of the following lend support to the notion that sleep is an adaptive response except:

    a. Large predators like lions can sleep whenever and wherever they want.
    b. Cows sleep very little because they need to remain awake to guard against predators.
    c. The Indus dolphin sleeps while it swims, a total of seven hours per day.
    d. Shrews that need to eat a lot sleep very little.

5. When sleep-deprived subjects are permitted to sleep normally, they

    a. regain most of the stage 1 sleep they lost.
    b. do not regain all the sleep they lost.

    c. experience a nearly equal percentage of recovery for all stages of sleep.
    d. go directly into REM sleep from waking.

6. The effects of forced exercise using the yoked-control technique

    a. increased subjects' needs for REM sleep on recovery nights following the experiment.
    b. reduced the total sleep time of experimental and control subjects by the same amount.
    c. include an increase in body temperature.
    d. are exaggerated when subjects are fed an enriched diet.

7. Exercise increases the amount of slow-wave sleep only if

    a. body temperature rises.
    b. brain temperature rises.
    c. body temperature decreases.
    d. brain temperature decreases.

8. Subjects were treated to an interesting outing before spending the night in a sleep laboratory in order to

    a. increase mental activity without affecting physical activity.
    b. eliminate affects of external stress.
    c. reduce fluctuations in metabolic rate.
    d. maintain a constant level of alertness.

9. Which of the following statements about thermoregulation and sleep is not true?

    a. Warm baths increase the amount of slow-wave sleep.
    b. People's alertness during the day is related to their body temperature.

c. The anterior hypothalamus and adjacent preoptic area contain neural circuits involved in thermoregulation.

d. Humans can save a considerable amount of energy by lowering body temperature during sleep.

10. The rebound phenomenon

a. indicates that REM sleep has the same function as slow-wave sleep.

b. occurs when REM sleep-deprived subjects are permitted to sleep normally.

c. results when aspects of REM sleep intrude into wakefulness.

d. suggests that REM sleep deprivation causes physiological harm.

11. The REM sleep of rats who were trained to run a complex maze

a. increased until the maze was mastered and then returned to baseline levels.

b. decreased until the maze was mastered and then returned to baseline levels.

c. increased each day during training.

d. decreased each day during training.

12. All of the following suggest that restoration and repair take place during slow-wave sleep except:

a. Slow-wave sleep increases following periods of emotional stress.

b. Growth hormone is secreted during slow-wave sleep.

c. Protein synthesis is increased during sleep, especially in animals like rats.

d. Cerebral blood flow is decreased during slow-wave sleep.

Lesson II: Physiological Mechanisms of Sleep, Sleep Disorders, and Biological Clocks

Read the interim summary on page 277 of your text to re-acquaint yourself with the material in this section.

Learning Objective 9-5   Evaluate evidence that the onset and amount of sleep is chemically controlled, and describe the neural control of arousal.

Read pages 266-270 and answer the following questions.

1. Cite evidence that indicates that sleep is regulated. (Karacan et al., 1970)

2. Study Figure 9.9 in your text and briefly explain how sleep might be triggered chemically by either a sleep-promoting substance or a wakefulness-promoting substance.

3. Review attempts to learn more about the control of sleep.

a. Review sleep patterns of the cerebral hemispheres of dolphins first discussed in Learning Objective 9-2. (Mukhametov, 1984)

b. Now explain what these results indicate about the chemical control of sleep and the circulation of these chemicals in the blood.

4. After an extensive literature review, what have Borbély and Tobler (1989) concluded about the chemical control of sleep?

5. Identify two categories of drugs that affect sleep and what their effects suggest about the chemical control of sleep.

6. Summarize the hypothesis of Benington et al. (1995) about what it means for the brain to become "tired."

7. Define *arousal* in your own words.

8. Electrical stimulation of the reticular formation produces arousal. (Moruzzi and Magoun, 1949. See Figure 9.10 in your text.)

   a. Where is the reticular formation located?

   b. How does it receive sensory information?

   c. How does sensory input affect the reticular formation?

   d. How does the reticular formation, in turn, affect the cerebral cortex? through which pathways? (Jones, 1990)

9. Which four systems of neurons play a role in arousal and wakefulness? (Marrocco et al., 1994)

10. Explain why the neurons of the locus coeruleus are affected by drugs like amphetamine. (To see the location of the locus coeruleus, study Figure 9.11 in your text.)

11. Review research by Aston-Jones and Bloom (1981a, 1981b, and 1994) on the activity patterns of noradrenergic neurons in the locus coeruleus.

    a. What did recordings of single neurons in freely moving cats indicate about the firing rate during wakefulness and REM sleep? (Study Figure 9.12 in your text. Aston-Jones and Bloom, 1981a).

    b. What do these differences in firing rate suggest about the control of arousal and REM sleep?

    c. When is the activity of noradrenergic LC neurons most sensitive to environmental stimuli? When is the firing rate low? (Aston-Jones and Bloom, 1981a, 1981b)

    d. Describe the task Aston-Jones et al. (1994) taught monkeys.

e. Study Figure 9.13 in your text and note when noradrenergic neurons responded vigorously and when they slowed down. How did this activity correspond to the monkeys' performance?

f. What conclusion do these results support?

12. Outline research that suggests that acetylcholine and serotonin, along with norepinephrine, play a role in the control of arousal.

a. What happens when acetylcholinergic neurons in the pons and basal forebrain are stimulated? (Jones, 1990; Steriade, 1996)

b. How do acetylcholinergic antagonists affect EEG signs of arousal? acetylcholinergic agonists? (Vanderwolf, 1992)

c. How did ACh levels in three brain regions associated with arousal correlate with the animals' level of activity? (Day et al., 1991)

d. Where are almost all of the brain's serotonergic neurons found? (Study Figure 9.14 in your text.)

e. Where do their axons project?

f. How is cortical arousal affected by stimulation of the raphe nuclei? administration of PCPA? (Peck and Vanderwolf, 1991)

g. Compare the response of noradrenergic and serotonergic neurons to external stimuli that produce pain or induce stress. (Jacobs et al., 1990)

13. Summarize the following researchers' suggestions about the role of serotonergic neurons.

a. Jacobs and Fornal (1993)

b. Marrocco, Witte, and Davidson (1994)

14. Study Figure 9.15 in your text and describe the changes in the firing rate of serotonergic neurons during waking and slow-wave and REM sleep. (Trulson and Jacobs, 1979)

15. To review: List the three systems of neurons that appear to play important roles in arousal and where in the brain they may exert their effects.

Neural System   Possible Function

1.

2.

3.

Learning Objective 9-6   Discuss the neural control of slow-wave and REM sleep.

Read pages 270-277 and answer the following questions.

1. a. How did destruction of the basal forebrain region affect the sleep and health of rats? (Nauta, 1946) of cats? (McGinty and Sterman, 1968)

    b. What other experimental procedures corroborated the results of lesion studies? (Szymusiak and McGinty, 1968b; Sterman and Clemente, 1962a, 1962b)

2. a. Neurons in the _____ and the _____ (often referred to as the _____) of the basal forebrain region are important in _____ regulation. Some of these neurons respond to changes in _____ temperature and others respond to information from thermosensors in the _____.

    b. If the POAH is warmed, what is the effect? (McGinty et al., 1994)

    c. What happens to the firing rate of neurons in the POAH when an animal falls asleep? When body temperature rises? (See Figure 9.16 in your text. Scammell et al., 1993; Alam et al., 1995a, 1995b)

    d. Cite other evidence that brain functioning is related to temperature. (Kolvalzon and Mukhametov, 1982; Hayaishi, 1988; Naito et al., 1988)

3. a. What is the earliest component of REM sleep recorded from laboratory animals?

    b. Describe PGO waves and explain the name. (See Figure 9.17 in your text.)

    c. Why can we only speculate that PGO waves occur in humans?

4. Why do people who have been exposed to organophosphate insecticides engage in more REM sleep? (Stoyva and Metcalf, 1968)

5. a. When are the levels of acetylcholine released by the terminal buttons in the cerebral cortex of the cat highest? lowest? (Jasper and Tessier, 1969)

    b. Use one word to describe the rate of glucose metabolism during REM sleep in brain regions that contain ACh-secreting neurons or receive input from them. (Lydic et al., 1991)

6. The _____ neurons play a central role in triggering the onset of REM sleep. They are located in the region which is referred to as the _____. (See Figure 9.18 for an illustration.)

7. a. Study Figure 9.19 in your text and describe the activity of the *REM-ON* cell.

    b. What might the increased activity of these acetylcholinergic cells signify?

8. What effect do peribrachial lesions have on REM sleep? (See Figure 9.20 in your text. Webster and Jones, 1988)

9. a. What happens to an animal's REM sleep when carbachol is infused into the region of the pons ventral to the locus coeruleus? (In your text, this region is referred to as the _____ _____

    _____ _____.)

    b. What are the effects of carbachol? (Quattrochi et al., 1989)

10. What is the effect of lesions of the MPRF on REM sleep? (See Figure 9.21 in your text. Siegel, 1989)

11. What parts of the brain are responsible for the following components of REM sleep?

    a. arousal and cortical desynchrony

    b. PGO waves (Sakai and Jouvet, 1980)

    c. rapid eye movements (Webster and Jones, 1988)

12. a. When Jouvet (1972) made a lesion caudal to the acetylcholinergic neurons in the dorsolateral pons of a laboratory cat, what behavioral change did he observe?

    b. What is the location of the neurons whose axons are responsible for the muscular paralysis of REM sleep?

    c. Where do these axons project caudally? (Sakai, 1980)

    d. And where do axons from the magnocellular nucleus project? What kind of synapses do they form here? (Morales et al., 1987)

13. Review research that supports the role of this pathway in REM sleep paralysis.

    a. How do lesions of the subcoerulear nucleus affect REM sleep and its accompanying paralysis? (Shouse and Siegel, 1992)

b. What happens to single neurons in the magnocellular nucleus during REM sleep? (Kanamori et al., 1980)

c. What is the effect of electrical stimulation of the magnocellular nucleus? (Sakai, 1980) lesions? (Schenkel and Siegel, 1989)

d. Identify the inhibitory transmitter substance found in the magnocellular nucleus and its likely function. (Fort et al., 1990)

14. What does this presence of this inhibitory mechanism suggest about the importance of the motor components of dreaming?

15. Study Figure 9.22 in your text and review the neural circuitry of REM sleep.

a. The first event preceding REM sleep is the activation of _____ neurons in the _____ area of the _____ _____.

b. The acetylcholinergic neurons directly activate brain stem mechanisms responsible for what phenomenon?

c. In order to produce atonia, the acetylcholinergic neurons activate neurons in what area? To what brain structure are these neurons connected?

d. Finally, the cortical activation that accompanies REM sleep is produced by neurons in the _____, that in turn activate acetylcholinergic neurons of the _____ _____.

16. Even though the MPRF is responsible for only one component of sleep, infusion of an ACh agonist induces REM sleep. Why might this be? (Reinoso Suàrez et al., 1994)

17. What activity might be responsible for the rebound effect after REM sleep deprivation? (Mallick et al., 1989)

18. The firing rates of _____ neurons of the locus coeruleus and the serotonergic neurons of the _____ _____ are lowest during _____.

19. What, then, may trigger a bout of REM sleep? (See Figures 9.23 and 9.24 in your text.) Review the experimental evidence.

Read the interim summary on page 280 of your text to re-acquaint yourself with the material in this

Learning Objective 9-7   Discuss insomnia, sleeping medications, and sleep apnea.

Read pages 277-278 and answer the following questions.

1. Describe insomnia, including its incidence, problems of definition, and most important cause.

2. Explain how the use of sleeping medication often leads to

    a. drug tolerance.

    b. a withdrawal effect. (Weitzman, 1981)

    c. drug dependency insomnia. (Kales et al., 1979)

3. What is the appropriate goal of sleeping medication?

4. a. Describe sleep apnea. Carefully explain the change in blood levels of carbon dioxide and its effect.

    b. What is a frequent cause of sleep apnea and how is it corrected? (Sher, 1990; Westbrook, 1990)

Learning Objective 9-8   Discuss problems associated with REM and slow-wave sleep.

Read pages 278-280 and answer the following questions.

1. Describe the four symptoms of narcolepsy.

    a. sleep attack

    b. cataplexy

    c. sleep paralysis

    d. hypnagogic hallucination

2. Describe the results of research on

    a. the sleep patterns of narcoleptic subjects. (Rechtschaffen et al., 1963)

    b. the heritability of narcolepsy. (Nishino et al., 1994, 1995; Nitz et al., 1995)

3. Which two kinds of drugs are used to treat narcolepsy? (Mitler, 1994; Hublin, 1996)

4. a. What happens when people suffering from REM without atonia dream? (Schenck et al., 1986)

    b. The symptoms of REM without atonia are the opposite of those of _____.

c. What happens to the motor cortex and subcortical motor systems during REM without atonia? What appears to be the cause?

5. Describe three maladaptive behaviors that may occur during slow-wave sleep, noting any association with sleep stages, who is most susceptible, and what the best treatments are.

Read the interim summary on pages 287-288 of your text to re-acquaint yourself with the material in this section.

> *Learning Objective 9-9*   Describe circadian rhythms; discuss research on the neural and physiological bases of biological clocks.

Read pages 280-287 and answer the following questions.

1. Define *circadian rhythms* in your own words.

2. Study the record of wheel-running activity of a rat under various conditions of illumination presented in Figure 9.26 in your text. Use this example to continue your explanation of daily rhythms.

    a. During which portion of a normal day/night cycle is the rat active?

    b. When the "day" was artificially advanced 6 hours, what happened to the rat's activity cycle?

    c. What effects did constant dim light have on the rat's activity?

    d. Because there were no stimuli in the rat's environment that varied throughout the day, what must have been the source of rhythmicity?

3. a. What term describes the effect of light on daily activity cycles?

    b. How do pulses of light affect the activity cycles of animals kept in constant darkness? (Aschoff, 1979)

4. Describe the circadian rhythm of a human under normal illumination and constant illumination.

5. Circadian rhythms are controlled by biological clocks. Name the structure that contains the primary biological clock of the rat. (Moore and Eichler, 1972; Stephan and Zucker, 1972.)

6. How do lesions of the SCN affect

    a. running, drinking, and hormonal secretions?

    b. the timing of sleep cycles? (Ibuka and Kawamura, 1975; Stephan and Nuñez, 1977)

c. total amount of sleep?

7. Let's look at the anatomy of the SCN more closely. Study Figure 9.27 in your text and locate the SCN.

   a. Approximately how big is the SCN of the rat (Meijer and Rietveld, 1989)?

   b. What is unique about the dendrites of neurons found here?

   c. What physical evidence suggests that a group of neurons near capillaries that serve the SCN may be neurosecretory cells? (Card et al., 1980; Moore et al., 1980)

   d. And what does the presence of these neurosecretory cells suggest about the way the SCN functions?

   e. What region projects fibers to the SCN? (Hendrickson et al., 1972) What is the name of the pathway? (Again, look at Figure 9.27.)

8. Why are the circadian rhythms of *rd/rd* mice particularly interesting?

9. a. What protein is produced when light pulses reset an animal's circadian rhythm?

   b. What does the presence of this protein further indicate about the effects of light? (Rusak et al., 1990, 1992)

   c. Look at Figure 9.28 in your text and describe the effect of drugs that block glutamate receptors. (Abe et al. 1991; Vindlacheruvu et al., 1992)

10. a. Besides the retinohypothalamic pathway, from what other part of the brain does the SCN receive information?

    b. What substance is released from the terminal buttons of the neurons that connect the IGL to the SCN?

    c. What is the role of the connection between the IGL and the SCN? What does damage to this pathway do? (Harrington and Rusak, 1986)

11. In addition to light, what other stimuli reset an animal's circadian rhythms?

12. To which parts of the brain do neurons of the SCN connect in order to control drinking, eating, sleeping, and hormone secretion?

13. a. How did Lehman et al., (1987) abolish the circadian rhythms of subject animals and then reestablish them?

    b. What do the studies on grafts suggest about the means by which the suprachiasmatic nucleus controls circadian rhythms?

14. a. Explain the experimental procedure that Schwartz and Gainer (1977) used to demonstrate day-night fluctuations in the metabolic activity of the SCN.

    b. Study Figure 9.29 in your text and explain the results.

15. Summarize research on the how the SCN keeps biological time.

    a. What does research on prenatal and postnatal development of the SCN of the rat suggest controls its internal rhythm? (Moore and Bernstein, 1989; Reppert and Schwartz, 1984)

    b. How does research using drug infusion support this hypothesis? (Schwartz et al., 1987)

    c. Study Figure 9.30 in your text and explain what research by Welsh et al.(1995) on rat SCN cells indicates.

16. What is the most likely explanation for the synchronized activity of SCN neurons?

17. Summarize the research on the intracellular ticking in the common fruit fly.

    a. Describe the cycle of the *per* and *tim* genes in your own words.

    b. What is the effect of light on *tim*?

    c. Study Figure 9.31 in your text and describe the effect of exposure to pulses of light on the fruit fly's circadian rhythms.

18. Review the physiological control of seasonal rhythms.

    a. What is the relationship between the male hamster's testosterone cycle and light? What effect do lesions of the SCN have on this cycle?

    b. What gland is involved in seasonal rhythms, and what hormone does it secrete? (See Figure 9.32 in your text.)

    c. Briefly describe how the SCN and this gland interact to control seasonal rhythms.

19. a. Identify two changes in circadian rhythms and the physiological mechanisms that control them.

    b. How can the adverse effects of these changes be improved?

Lesson II Self Test

1. Which neurotransmitter does not play an excitatory role in arousal or wakefulness?

    a. acetylcholine
    b. norepinephrine
    c. neuropeptide Y
    d. serotonin

2. The firing rate of noradrenergic neurons in the locus coeruleus correlates best with

    a. the activity of neurons in the SCN.
    b. vigilance.
    c. the activity of ACh neurons.
    d. the basic rest activity cycle.

3. Which statement is not true of the POAH?

    a. This part of the brain is involved in thermoregulation.
    b. The POAH is part of the basal forebrain; the preoptic area and the adjacent anterior hypothalamus.
    c. Warming the POAH can induce slow-wave sleep.
    d. Neurons in the POAH are directly sensitive to changes in brain temperature, however they are not sensitive to thermosensors in the skin.

4. PGO waves

    a. are brief bursts of electrical activity that originate in the basal forebrain.
    b. are the first sign of a bout of REM sleep in animals.
    c. control muscular paralysis.
    d. are detected by an enzyme stain.

5. All of the following are involved in the neural circuitry of REM sleep except:

    a. raphe nuclei
    b. peribrachial area
    c. acetylcholinergic neurons
    d. medial pontine reticular formation

6. The right amount of sleep is

    a. only obtained by infants.
    b. whatever seems to be enough.
    c. assured with sleeping medication.
    d. infrequently obtained by insomniacs.

7. Sleep apnea is

    a. a form of pseudoinsomnia.
    b. a side affect of sleeping medications.
    c. a period of sleep without dreams.
    d. the inability to sleep and breathe at the same time.

8. During a cataplectic attack, the individual

    a. awakens gasping for breath.
    b. tries to act out dreams.
    c. is unconscious.
    d. is overcome by muscular paralysis.

9. Which of the following is not true about biological clocks?

    a. They control circadian and seasonal rhythms.
    b. They include structures such as the SCN and pineal gland.
    c. They are often synchronized by zeitgebers.
    d. They are found only in mammals.

10. How did the daily behavior of a rat change with constant dim illumination?

    a. Periods of activity increased and periods of sleep decreased.
    b. Food consumption increased with activity levels.
    c. The biological clock ran slower; activity began about one hour later each day.
    d. Body temperature increased slightly with constant light.

11. The primary biological clock of the rat is located in the

    a. suprachiasmatic nucleus.
    b. reticular formation.
    c. locus coeruleus.
    d. pons.

12. All of the following are true about the "ticking" of the biological clock, except:

    a. "Ticking" is intrinsic to individual neurons.
    b. Infusion of the drug TTX abolishes both circadian rhythms and "ticking".
    c. activity cycles of neurons in the SCN are synchronized.
    d. The per and tim genes appear to control intracellular ticking.

Answers for Self Tests

	Lesson I		
1.	b	Obj. 9-1	
2.	a	Obj. 9-1	
3.	a	Obj. 9-1	
4.	c	Obj. 9-2	
5.	b	Obj. 9-3	
6.	c	Obj. 9-3	
7.	b	Obj. 9-3	
8.	a	Obj. 9-3	
9.	d	Obj. 9-3	
10.	b	Obj. 9-4	
11.	a	Obj. 9-4	
12.	a	Obj. 9-4	

	Lesson II		
1.	c	Obj. 9-5	
2.	b	Obj. 9-5	
3.	d	Obj. 9-6	
4.	b	Obj. 9-6	
5.	a	Obj. 9-6	
6.	b	Obj. 9-7	
7.	d	Obj. 9-7	
8.	d	Obj. 9-8	
9.	d	Obj. 9-9	
10.	c	Obj. 9-9	
11.	a	Obj. 9-9	
12.	b	Obj. 9-9	

CHAPTER 10
Reproductive Behavior

Lesson I: Sexual Development and Hormonal Control of Sexual Behavior

Read the interim summary on pages 296-297 of your text to re-acquaint yourself with the material in this section.

Learning Objective 10-1   Describe mammalian sexual development and explain the factors that control it.

Read pages 290-296 and answer the following questions.

1. Define *sexually dimorphic behaviors* in your own words and list several examples.

2. a. How many pairs of chromosomes do cells other than sperms and ova contain?

    b. Explain why gametes contain only one half of each pair of chromosomes.

3. How many types of sex chromosomes are there? What is the sex chromosome pattern for a male? for a female?

4. a. At what point is the sex of an offspring determined?

    b. Which parent determines the sex of an offspring? (See Figure 10.1 in your text.)

5. What event is responsible for sexual dimorphism?

6. a. List the three categories of sex organs.

     1.   2.   3.

    b. Now list the gonads.

     1.   2.

    c. What are their dual functions?

d. What factor determines whether the initially identical gonads become ovaries or testes? Cite research to support your answer. (Smith, 1994; Koopman et al., 1991)

7. Explain the difference between organizational effects and activational effects of sex hormones.

8. Name the precursors of the male and female internal sex organs. (Study Figure 10.2 in your text to see how the precursors of the internal sex organs continue to develop.)

9. a. Name the two types of hormones secreted by the testes that determine whether the Müllerian system or the Wolffian system continues to develop and describe their effects.

   b. Name the specific androgens responsible for masculinization.

10. Be sure that you understand how hormones exert their effect on the body and then explain the cause and consequences of

    1. androgen insensitivity syndrome. (See Figure 10.3 in your text.)

    2. persistent Müllerian duct syndrome.

11. What do people with Turner's syndrome reveal about the hormones necessary for the development of female internal and external sex organs? Explain your answer. (See Figure 10.4 in your text.)

12. Stop now and review what you have learned. Cover the right half of Figure 10.5 in your text (including the triple arrows). Identify the hormones that are secreted by the newly developed testes or ovaries and describe their effect on the development of the internal sex organs and external genitalia.

13. Complete these statements. (See Figure 10.6 in your text.)

    a. The primary sex characteristics include

    b. The secondary sex characteristics include

    c. Puberty begins when cells in the hypothalamus

    d. Gonadotropin-releasing hormones in turn stimulate

    e. The two gonadotropic hormones are

    f. Although the gonadotropic hormones are named for the effects they produce in the female,

g. The steroid sex hormones important for sexual maturation in the female are _____, an

_____, and in the male _____, an _____.

14. Summarize the changes in the bodies of males and females that are initiated by gonadal hormones at puberty.

15. What evidence confirms that the bipotentiality of some secondary sex characteristics is lifelong.

Table 10.1 in your text summarizes information about sex steroid hormones.

Read the interim summary on pages 310-311 of your text to re-acquaint yourself with the material in this section.

Learning Objective 10-2   Describe the hormonal control of the female reproductive cycle and of male and female sexual behavior.

Read pages 297-301 and answer the following questions.

1. What is the chief difference between the human menstrual cycle and the estrous cycle of other female mammals?

2. Study Figure 10.7 in your text and follow the menstrual cycle from beginning to end.

   a. Name the principal hormone that stimulates the growth of ovarian follicles.

   b. Name the hormone secreted by the maturing ovarian follicle.

   c. What changes begin to occur in the uterus in response to this hormone?

   d. What effect does this hormone have on the anterior pituitary gland?

   e. What does the hormone from the anterior pituitary gland do?

   f. After the ovum is released, what happens to the ruptured follicle?

   g. What hormones does this structure release?

   h. If the ovum is not fertilized or is fertilized too late, what changes occur?

3. List the three features common to all male sexual behavior.

   1.   2.   3.

4. What is the refractory period?

5. What is the Coolidge effect and what may be its evolutionary importance?

6. What evidence indicates the importance of testosterone in male sexual behavior? (Bermant and Davidson, 1974)

7. a. Through aromatization _____ is converted to _____. Briefly explain this process illustrated in Figure 10.8 in your text.)

b. What happens to sexual behavior of adult male monkeys if they are given a drug that blocks aromatase? (Zumpe et al., 1993)

8. a. Where is oxytocin produced and when is it released in both males and females?

    b. What are the effects of its release on females? on males? (Carter, 1992c; Carmichael et al., 1994)

9. a. Where is prolactin produced and when is it released in males? (Oaknin et al., 1989)

    b. What is its effect on male sexual behavior? (Foster et al., 1990) Cite research to support your answer. (Doherty et al., 1986; Mas et al., 1995)

10. a. Name the position that receptive females of many four-legged species will assume to facilitate copulation.

    b. What two hormones are required for the sexual response of a female rodent?

    c. Which hormone "primes" the other? Explain. (Takahashi, 1990)

11. List and briefly explain the effects that the sequence of estradiol followed by progesterone have on the female rat.

    1.

    2.

    3.

12. Define *behavioral defeminization* and *behavioral masculinization* in your own words.

13. Using these terms, explain the effects of the following treatments on sexual behavior:

    a. an adult male, castrated at birth and given no hormones, then given estradiol and progesterone in adulthood (Blaustein and Olster, 1989)

    b. an adult male, castrated in adulthood, then given estradiol and progesterone in adulthood

    c. an adult female, ovariectomized at birth and given testosterone, then given estradiol and progesterone in adulthood

    d. an adult female, ovariectomized at birth and given testosterone, then given testosterone in adulthood (Breedlove, 1992,; Carter, 1992b)

14. Check your understanding of the organizational effects of androgens by studying Figure 10.9 in your text and then completing the blanks in Table 1, below.

   a. Circle the entry that indicates the activational effect of estradiol and progesterone. Label it AE.

   b. Circle the entry that indicates evidence of behavioral defeminization. Label it BDF.

   c. Circle the entry that indicates evidence of behavioral masculinization. Label it BM.

Hormone Treatment		Resulting Sexual Behavior	
Immediately after birth	*When rat is fully grown*		
None	Estradiol + progesterone	Female: _____	Male: _____
None	Testosterone	Female: _____	Male: _____
Testosterone	Estradiol + progesterone	Female: _____	Male: _____
Testosterone	Testosterone	Female: _____	Male: _____

Table 1

15. Outline two explanations why all fetuses, both male and female, do not become masculinized and defeminized from exposure to estrogens. (Breedlove, 1992)

16. a. Female rats lick the genital region of their offspring. Why is this a useful behavior for both mother and pups?

   b. Why do rat mothers spend more time licking their male offspring? (reviewed by Moore, 1986)

   c. When the mothers' ability to smell was destroyed, how did their behavior toward their male offspring change?

   d. How was the adult sexual behavior of these males affected by reduced attention from their mothers? by genital stroking by the researchers?

   e. What do these results suggest about some of the masculinizing effects of androgens?

Learning Objective 10-3 Describe the role of pheromones in reproductive and sexual behavior.

Read pages 301-304 and answer the following questions.

1. Pheromones transmit chemical messages from one _____ to another, unlike hormones, which transmit messages from one part of the _____ to another. Pheromones are usually detected through _____. They can affect reproductive _____ or _____.

2. Describe each of the following phenomena which affect reproductive physiology.

    a. Lee-Boot effect (van der Lee and Boot, 1955)

    b. Whitten effect (Whitten, 1959)

    c. Vandenbergh effect (Vandenbergh et al., 1975)

    d. Bruce effect (Bruce, 1960a, 1960b)

3. a. Name and describe the location of the sensory organ that mediates the effects of pheromones.

    b. Where do afferent axons from this organ project? (Wysocki, 1979. See Figure 10.10 in your text.)

    c. What kind of compounds does the vomeronasal organ most likely detect? Cite research to support your answer. (Meredith and O'Connell, 1979; Meredith, 1994)

    d. Cite evidence that the accessory olfactory bulb is essential for pheromone detection. (Halpern, 1987)

4. Trace the neural circuit responsible for the effects of these pheromones beginning with the region to which the accessory olfactory bulb projects. (Study Figure 10.11 in your text.)

5. a. If a female mouse mates with a male and later encounters him, will his odor cause her to abort?

    b. Which axons appear to play a role in learning to identify a particular odor?

    c. Following destruction of these axons, how well did the female recognize the odor of her mate? What was the subsequent effect on her pregnancy? (Keverne and de la Riva, 1982)

6. Review research to understand why pregnant females do not abort if they later encounter the odor of the male with which they mated.

    a. How are the noradrenergic axons affected by vaginal stimulation? (Rosser and Keverne, 1985)

    b. What neurotransmitter is necessary for olfactory learning to occur? (Gray et al., 1986; Leon, 1987)

7. a. By what two means, are male hamsters exposed to pheromones present in the vaginal secretions of females?

b. What behavior do these pheromones elicit?

c. Under what circumstances is this mating behavior abolished? (Powers and Winans, 1975; Winans and Powers, 1977; Lehman and Winans, 1982)

d. If aphrodisin, a sex-attractant pheromone of female hamsters, is swabbed on the hindquarters of a male hamster, how do test males react? (Singer et al., 1986; Singer, 1991)

8. a. How does destruction of the vomeronasal organ of female rats affect their

1. preference for normal males and castrated males? (Romero et al., 1990)

2. sexual receptivity? (Rajendren et al., 1990)

b. What do these results suggest about the production of sex-attractant pheromones by males?

9. Discuss the affects of pheromones in vaginal secretions or sweat in

a. synchronizing women's menstrual cycles. (McClintock, 1971; Russell et al., 1977)

b. female sexual attractiveness. (Doty et al., 1975)

c. female subjects' tendency to interact with men. (Crowley and Brookshank, 1991)

10. a. What did the examination of the olfactory mucosae of surgical patients reveal? (Garcia-Velasco and Mondragon, 1991)

b. What were the results of research to determine whether particular chemicals serve as pheromones in humans? Monti-Bloch et al., 1994)

Learning Objective 10-4 Discuss the activational effects of gonadal hormones on the sexual behavior of women and men.

Read pages 304-306 and answer the following questions.

1. Ovarian hormones control both the _____ and the _____ to mate of most estrous female mammals other than primates. (Wallen, 1990)

2. Compare the sexual receptivity of women throughout the menstrual cycle with that of other female mammals throughout the estrous cycle.

3. a. What have most studies of the influence of ovarian hormones on women's sexual interest concluded? (Adams et al., 1978; Morris et al., 1987)

    b. What did Wallen (1990) point out about the sexual interest and sexual activity of women in these studies?

    c. When does sexual activity of lesbian couples tend to increase and what does this pattern suggest about the role of ovarian hormones? (Matteo and Rissman, 1984)

4. Now compare the sexual activity across the menstrual cycle of small numbers of monkeys living in small cages with that of large numbers of monkeys living in a large cage. (Study Figure 10.12 in your text. Wallen et al., 1986)

5. How do oral contraceptives affect fluctuations in a woman's secretion of ovarian hormones and her sexual interest? (Alexander et al., 1990)

6. Describe research on the activational effects of androgens on female sexual behavior in human couples.

    a. frequency of intercourse and female sexual gratification (Persky et al., 1978; Morris et al., 1987)

    b. female sexual desire (Alexander and Sherwin, 1993; Sherwin et al., 1985; Sherwin, 1994)

7. What roles, still speculative, may oxytocin play in female sexual response?

8. Compare the activational effects of testosterone on the sexual behavior of human males and other male mammals.

9. a. If a man is castrated, what changes occur in his interest in sexual activity?

    b. What factors appear to influence the decline in sexual ability after castration? (Money and Ehrhardt, 1972)

    c. Cite research with cats that also suggests the importance of experience.(Rosenblatt and Aronson, 1958a, 1958b)

10. Describe research that confirms that testosterone levels are affected by sexual arousal.

    a. the beard growth of an isolated scientist (Anonymous, 1970)

    b. watching erotic films (Hellhammer et al., 1985)

11. What roles, again still speculative, may oxytocin and prolactin play in male sexual response?

Learning Objective 10-5   Discuss sexual orientation, the prenatal androgenization of genetic females, and the failure of androgenization of genetic males.

Read pages 306-310 and answer the following questions.

1.   Explain sexual orientation in your own words.

2.   _____ homosexuality occurs only in humans. (Ehrhardt and Meyer-Bahlberg, 1981)

3.   a. What belief concerning the cause of homosexuality was dispelled by a large-scale study by Bell et al., (1981)?

     b. What factor did they find to be the best predictor of homosexuality?

     c. What do these results suggest is a more likely cause for homosexuality than childhood social interactions?

4.   a. Compare the levels of sex hormones during adulthood of male and female homosexuals with heterosexual males and females (Meyer-Bahlberg, 1984)

     b. What, then, can we conclude about the biological cause of homosexuality?

     c. What is a more likely biological cause?

5.   a. What is the cause of congenital adrenal hyperplasia (CAH)?

     b. How does prenatal masculinization affect males and females fetuses?

     c. Once diagnosed, how are human females medically treated?

     d. What do females with this syndrome report about their sexual orientation? (Money et al., 1984)

     e. Outline an explanation of how abnormally high levels of androgens may exert a behavioral effect.

6.   Describe how Goy et al., (1988) studied the lasting effects of prenatal androgenization in female monkeys.

     a. How did the researchers create the conditions for prenatal androgenization?

b. What variable influenced the degree of genital masculinization?

c. Compare the sociosexual behavior of normal female monkeys and androgenized female monkeys with normal genitals.

d. What do these results suggest?

7. a. Briefly describe how the internal and external genitalia of males with androgen insensitivity syndrome develop.

b. What is the best treatment of this condition and how is this confirmed at the time of puberty?

c. What is the sexual orientation and behavior of these adults? (Money and Ehrhardt, 1972)

d. Carefully explain what this syndrome suggests about the role of testosterone and aromatized testosterone in normal male development.

8. Briefly summarize some of the documented differences between men's and women's brains. (Breedlove, 1994; Swaab et al., 1995 for specific references.)

a. shared functions of the hemispheres

b. overall brain size

c. size of particular regions

9. What do most researchers believe accounts for the sexual dimorphism of the human brain?

10. a. Identify the three subregions of the brain that differ in size in heterosexual and homosexual men and heterosexual women. (Swaab and Hofman, 1990; LeVay, 1991; Allen and Gorski, 1992)

b. Compare the size of the bed nucleus of the stria terminalis (BNST) of

1. males and females
2. male transsexuals and females
3. male homosexuals and male heterosexuals.

c. Explain why the size of the BNST appears to be related to sexual identity and not sexual orientation by referring to these size comparisons. (Zhou et al., 1995; Study Figure 10.13 in your text.)

d. What limited conclusions can we draw from the results of these studies?

11. a. What event could interfere with the prenatal androgenization of males?

   b. Explain how maternal stress during pregnancy may have affected the

      1. sexual behavior of male rats (Ward, 1972).

      2. play behavior of juvenile male rats (Ward and Stehm, 1991).

      3. brain development of male rats (Anderson, et al., 1986).

   c. What do these results suggest about a biological cause of male homosexuality?

12. What were the results of twin studies of male and female homosexuality? (Bailey and Pillard, 1991; Bailey et al., 1993; Pattatucci and Hamer, 1995)

13. List the two biological factors that research indicates may affect sexual orientation.

   1.   2.

Lesson I Self Test

1. Which example illustrates the activational effects of sex hormones?

   a. development of ovaries and uterus
   b. production of sperm
   c. differentiation of the primordial gonads
   d. changes in brain development caused by androgens

2. The precursor of the _____ sex organs is the _____ system which develops _____.

   a. female; Wolffian; without any hormonal stimuli
   b. male; Wolffian; only if the testes secrete the appropriate hormones
   c. male; Müllerian; without any hormonal stimuli
   d. female; Müllerian; only if the ovaries secrete the appropriate hormones

3. Which event first marks the beginning of puberty?

   a. release of gonadotropic hormones by anterior pituitary gland
   b. secretion of gonadotropin-releasing hormones by hypothalamus

   c. appearance of secondary sex characteristics
   d. production of estrogens by ovaries or androgens by testes

4. The LH surge causes

   a. estrus.
   b. the refractory period.
   c. the release of milk.
   d. ovulation.

5. During aromatization _____ is converted into _____.

   a. prolactin; oxytocin
   b. aromatase; testosterone
   c. estrogen; estradiol
   d. testosterone; estradiol

6. The sexual behavior of female rodents depends on the presence of _____.

   a. testosterone
   b. estrogen
   c. prolactin and oxytocin
   d. estradiol and progesterone

7. The acceleration of the onset of puberty in a female rodent caused by the odor of a male is known as the _____ effect.

   a. Whitten
   b. Bruce
   c. Vandenbergh
   d. Lee-Boot

8. Removal of the _____ disrupts the Lee-Boot, Bruce, Vandenbergh, and Whitten effects.

   a. pituitary gland
   b. ventromedial nucleus of the hypothalamus
   c. accessory olfactory bulb
   d. adrenal glands

9. The frequency of intercourse throughout the menstrual cycle is at least moderately correlated with the woman's peak level of

   a. testosterone.
   b. progesterone.
   c. estradiol.
   d. oxytocin.

10. Castrated male cats remained potent longer if they

    a. were introduced to a different female each time.

   b. had previously engaged in high levels of sexual activity.
   c. were housed with intact males.
   d. did not have to compete with other males for the right to mate.

11. A large-scale study of male and female homosexuals found that

   a. homosexuality results from unhappy parent-child relationships.
   b. self-report was the best predictor of adult homosexuality.
   c. only children were more likely to be homosexual than children with siblings.
   d. homosexuality is often the result of poor interpersonal relationships with peers.

12. The biological basis of homosexuality may be differences in the

   a. organizational affects of prenatal hormones.
   b. activational affects of prenatal hormones.
   c. degree of sexual dimorphism of the prenatal brain.
   d. hormone levels of adult heterosexuals and homosexuals.

Lesson II:　Neural Control of Sexual Behavior and Parental Behavior

Read the interim summary on pages 316-317 of your text to re-acquaint yourself with the material in this section.

> *Learning Objective 10-6* Discuss the neural control of male sexual behavior.

Read pages 311-314 and answer the following questions.

1. a. Let's look first at the spinal mechanisms that play a role in male sexual behavior. List the two male sexual reflexes that are controlled by spinal mechanisms.

   1.   2.

   b. Explain how men with spinal cord damage may still become fathers. (Hart, 1978) may still experience orgasm. (Money, 1960; Comarr, 1970)

2. Why is the spinal nucleus of the bulbocavernosus (SNB) larger in male rats than female rats? (Breedlove and Arnold, 1980, 1983; Arnold and Jordan, 1988)

3. a. What stimulates a mother rat to lick the anogenital region of her male pups?

    b. If the mother's sense of smell is destroyed and the amount of licking diminishes, what structural change results in the brains of the male pups? (Moore et al., 1992)

4. Now let's look at the brain mechanisms that play a role in male sexual behavior. Name and describe the location of the brain region most important for male sexual behavior.

5. a. If the medial preoptic area (MPA) is electrically stimulated, what kind of behavior is elicited in a male rat? (Malsbury, 1971)

    b. How does sexual activity affect the

        1. electrical activity of the neurons in the MPA? (Shimura et al., 1994; Mas, 1995)

        2. metabolic activity of a male's MPA? (Oaknin et al., 1989; Robertson et al., 1991; Wood and Newman, 1993)

    c. What profound effect does the destruction of the MPA have on male sexual behavior? (Heimer and Larsson, 1966/1967)

6. How does the MPA receive chemosensory input? somatosensory input? (See Figure 10.14 in your text.)

7. a. Where is the sexually-dimorphic nucleus (SDN) located? (Gorski et al., 1978)

    b. What factor during prenatal development determines the size of the sexually dimorphic nucleus?

8. a. How is the volume of the SDN of an individual male rat related to the animal's sexual activity? (Anderson et al., 1986)

    b. And how do lesions of the SDN affect male sexual activity? (De Jonge et al., 1989)

9. Explain how the sexual function of a male rodent, castrated in adulthood, can be restored and why. Be sure to refer to the activational effects of hormones. (Davidson, 1980; Nyby et al., 1992; Roselli et al., 1989)

10. a. How is the sexual behavior of male monkeys affected by damage to the medial preoptic area? (Slimp et al., 1975)

    b. And what change occurs in some neurons in the MPA just before an animal begins pursuing a female? (Shimura et al., 1994)

c. What do the results of these studies suggest about the role of the MPA in sexual behavior?

11. a. Where do the axons of neurons in the MPA project?

b. How does destruction of these axons affect male sexual behavior? (Brackett and Edwards, 1984)

c. In general, how did copulatory behavior affect the activity of neurons in this region? (Shimura and Shimokochi, 1990)

d. What do all these results suggest is the means by which the MPA affects sexual behavior?

12. Summarize some results of research on the role that other parts of the brain may play in sexual behavior.

a. response to pheromones

b. copulatory behavior (DeJonge et al., 1992)

c. Fos protein production (Wood and Newman, 1993)

13. a. Describe how seizures originating in the temporal lobes sometime affect the sexual activity of human males. (Blumer and Walker, 1975; Morrell, 1991; Morrell et al., 1994)

b. Describe treatments that usually restores normal sexual drive.

14. Finally let's look at the role that neurotransmitters and peptide hormones may play in male sexual behavior.

a. What aspects of male sexual behavior may be influenced by oxytocin? (Arletti et al., 1992; Argiolas and Gessa, 1991)

b. How do injections of a vasopressin antagonist affect the sexual behavior of male rats? (Argiolas et al., 1988)

15. a. How does castration affect brain levels of vasopressin? (See Figure 10.16 in your text. DeVries, 1990)

b. If these castrated male rats are given drugs that stimulate vasopressin receptors, how is sexual behavior affected?

c. If testosterone is given to males several weeks after castration, how are vasopressin levels and sexual activity affected?

16. Finally, how do microinjections of a dopamine antagonist into the MPA affect sexual activity? a dopamine agonist? (Hull et al., 1986; Warner et al., 1991)

To review: Study Figure 10.17 in your text which summarizes this information.

Learning Objective 10-7 Discuss the neural control of female sexual behavior.

Read pages 315-316 and answer the following questions.

1. The most important forebrain region for male sexual behavior is the _____ _____

    _____ and the most important forebrain region for female sexual behavior is the _____

    _____ of the _____. (The location is shown in Figure 10.18 in your text.)

2. What are the behavioral effects of

    a. bilateral lesions of the ventromedial nuclei in female rats?

    b. electrical stimulation of the VMH? (Pfaff and Sakuma, 1979)

    c. injections of estradiol and progesterone in the VMH of ovariectomized females? (Rubin and Barfield, 1980; Pleim and Barfield, 1988)

    d. injections of a chemical that blocks the production of progesterone receptors? (Ogawa et al., 1994)

3. How does production of Fos protein in the VMH change with copulation or mechanical stimulation of the genitals or flanks? (Pfaus et al., 1993; Tetel et al., 1993)

4. a. Let's look at the mechanisms responsible for these behaviors. When female hamsters are given injections of estradiol and then progesterone, what change occurs in the activity of neurons in the VMH? (Rose, 1990)

    b. What kind of receptors are found on neurons in the VMH that increase Fos production in response to genital stimulation? (Tetel et al., 1994)

    c. Where, then, do the effects of estradiol and stimulation, converge?

    d. Explain how estradiol increases the effectiveness of progesterone and affects female sexual behavior. Cite research to support your answer. (Study Figure 10.19 in your text. Blaustein and Feder, 1979)

5. a. Where do axons of neurons in the VMH project?

    b. Summarize changes in female sexual behavior resulting from

        1. electrical stimulation of the PAG and lesions of the PAG? (Sakuma and Pfaff, 1979a, 1979b)

        2. lesions that disconnect the VMH from the PAG? (Hennessey et al., 1990)

        3. estradiol treatment or electrical stimulation of the VMH. (Sakuma and Pfaff, 1980a, 1980b)

    c. What do these results suggest about a role of the PAG in female sexual behavior?

6. a. Where do axons of neurons in the PAG project? Where do neurons there project?

b. What may be the function of this pathway?

7. Now summarize the results of some research on the role of neurotransmitters in female sexual behavior.

a. If female rats, primed with estradiol, receive progesterone, what physical change occurs in the VMH? (Schemata et al., 1990)

b. Under what circumstances do injections of oxytocin facilitate lordosis? (Schemata et al., 1989)

c. If female rats receive injections of an oxytocin antagonist, what behavioral change occurs? (McCarthy et al., 1994)

d. How does genital stimulation affect the activity of noradrenergic neurons? (Crowley et al., 1977)

e. If noradrenergic axons projecting to the spinal cord or forebrain are damaged, what behavioral change occurs? (Hansen et al., 1980; Hansen and Ross, 1983)

f. And how do injections of a norepinephrine agonist and antagonist into the hypothalamus affect estrous behavior? (Crowley et al., 1978; Fernandez-Guasti et al., 1985)

g. If a female rat already primed with estradiol and progesterone encounters a male, what happens to norepinephrine secretion in a limited region of the hypothalamus? (Study Figure 10.20 in your text. Vathy and Etgen, 1989)

Read the interim summary on pages 322-323 of your text to re-acquaint yourself with the material in this section.

Learning Objective 10-8   Describe the maternal behavior of rodents and explain how it is elicited and maintained.

Read pages 317-320 and answer the following questions.

1. Why is the attentive maternal behavior of a female rodent necessary for the survival of her offspring?

2. Describe the maternal behavior of a female rodent

a. during pregnancy. (See Figure 10.21 in your text.)

b. at the time of parturition.

3. Following birth, how does the mother

a. assist elimination? Be sure to explain the procedure Friedman and Bruno (1976) used to determine the mutually beneficial nature of this behavior.

b. retrieve pups outside the nest? (See Figure 10.22 in your text.)

4. When does maternal behavior begin to decrease?

5. Offer two reasons why maternal behavior is somewhat different from other sexually dimorphic behaviors.

6. a. Describe how a virgin female rat normally responds when she encounters a rat pup.

   b. Briefly explain how Fleming and Rosenblatt (1974) demonstrated the basis for this behavior.

7. a. Explain what the sensitization of virgin female rats means in your own words.

   b. Describe evidence that confirms the role of olfaction in sensitization. (Fleming and Rosenblatt, 1974; Fleming et al., 1979)

   c. In addition to olfaction, what physical sensations at the time of parturition also plays an important role? Cite research to support your answer. (Graber and Kristal, 1977; Yeo and Keverne, 1986)

8. Describe two situations that cause mouse pups to emit ultrasonic calls and indicate how female mice respond to these calls. (Noirot, 1972; Hofer and Shair, 1993; Ihnat et al., 1995)

9. To illustrate the importance of tactile stimulation in the maintenance of maternal behavior, explain behavioral changes that occur if the region around the mouth of the mother is desensitized. the mouths of the pups.

Learning Objective 10-9 Explain the hormonal and neural mechanisms that control maternal behavior and the neural control of paternal behavior.

Read pages 320-322 and answer the following questions.

1. Identify the hormones that facilitate nest building in nonpregnant female mice. (Lisk et al., 1969; Voci and Carlson, 1973)

2. Explain how the blood levels of these three hormones change with insemination, pregnancy, and parturition. (Study Figure 10.23 in your text.)

   a. estradiol

   b. progesterone

   c. prolactin

3. a. If ovariectomized virgin rats are given estradiol and progesterone in the pattern that duplicates the normal sequence, what behavioral change occurs? (Moltz et al., 1970; Bridges, 1984)

    b. What effects do these hormones have on a rat's responses to

       1. novel odors? (Fleming et al., 1989)

       2. preference for bedding material from a lactating female and her pups?

       3. the strength of the long-term effects of sensitization? (Fleming and Sarker, 1990)

4. Describe research on the possible interaction of prolactin with progesterone and estradiol. (Bridges et al., 1990)

5. a. Name the brain region critical for maternal behavior in rodents.
    b. How do lesions of the MPA affect maternal behavior and female sexual behavior? (Numan, 1974)

    c. What change occurs in the MPA as a result of

       1. parturition? (Del Cerro et al., 1995)

       2. exposure to pups?

    d. Describe the pathway critical for normal maternal behavior. Cite supporting research. (Numan and Smith, 1984; Numan and Numan, 1991)

    e. What kind of hormone receptors are found in the medial preoptic area? (Pfaff and Keiner, 1973)

    f. How does pregnancy affect these receptors? (Giordano et al., 1989)

    g. What is the effect on maternal behavior of estradiol implants in the MPA. (Numan et al., 1977)? antiestrogen chemical injections in the MPA. (Adieh et al., 1987)?

    h. Study Figure 10.24 in your text and describe how estradiol may exert its effects on maternal behavior.

6. Finally let's look at some research on the neural control of parental behavior.

    a. Compare the parental behavior and the MPA of monogamous prairie voles with that of promiscuous montane voles. (Shapiro et al., 1991)

b. If male prairie voles are exposed to pups, what change occurs in the MPA and what does this change suggest about the role of the MPA? (Kirkpatrick et al., 1994)

c. How do lesions of the MPA of ring doves affect paternal behavior induced by prolactin? (Slawski and Buntin, 1995)

d. How does mating or the birth of pups affect the secretion of vasopressin in the brains of male prairie voles? meadow voles? (Bamshad et al., 1993, 1994; Wang et al., 1994)

e. If sexually naive male prairie voles receive injections of vasopressin, what behavioral changes occur? How can these changes be abolished? (Study Figure 10.25 in your text. Wang et al., 1994)

Lesson II Self Test

1. Human males with spinal cord damage

   a. never again have an erection or experience an orgasm.
   b. can only experience an orgasm through mechanical stimulation.
   c. can experience "phantom erections" and orgasms.
   d. experience a decline in the ability to have an erection and orgasm similar to the effects of castration.

2. The size of the sexually dimorphic nucleus of the _____ is _____.

   a. medial amygdala; reduced in pups whose mother was prenatally sensitized
   b. left temporal lobe; directly related to level of prenatal stress
   c. preoptic area; controlled by the amount of androgens present during fetal development
   d. ventral tegmental area; directly related to fertility

3. Injections of oxytocin into the brains of male rats

   a. increase the likelihood of erection.
   b. appear to affect sexual motivation.
   c. decrease mounting behavior.
   d. reduce the number of successful intromissions.

4. The brain region most critical for female sexual behavior is the _____ and the brain region most critical for maternal behavior is the _____.

   a. ventromedial nucleus of the hypothalamus; ventral tegmental area
   b. medial preoptic area; sexually dimorphic nucleus
   c. ventromedial nucleus of the hypothalamus; medial preoptic area
   d. sexually dimorphic nucleus; periaqueductal gray matter

5. The priming effect of estradiol is caused by

   a. the LH surge.
   b. an increase in progesterone receptors.
   c. increased release of norepinephrine in the hypothalamus.
   d. an increase in the firing rate of neurons in the periaqueductal gray matter.

6. Electrical stimulation of the periaqueductal gray matter facilitates

   a. ovulation.
   b. lordosis.
   c. hormonal priming.
   d. lactation.

7. Female rodents lick the anogenital region of their young. What is one of the results of this behavior?

   a. assures identification of pups by scent
   b. recycles water
   c. reduces detection of pups' odor by intruders
   d. regulates pups' metabolic rate

8. Virgin female rats can be made to care for infants if they are

a. caged with an experienced mother.
b. allowed to observe pups through a glass partition.
c. given injections of progesterone.
d. placed with young pups for several days.

9. Ultrasonic calls from a rodent pups signals that the pup is

a. cold.
b. hungry.
c. threatened by an intruder.
d. unable to move.

10. If ovariectomized virgin female rats are given a sequence of doses of estradiol and progesterone, they will

a. build brood nests.
b. begin to lactate.
c. fail to retrieve pups.
d. be sensitized to care for young more quickly.

11. Just before parturition the level of estradiol

a. rises; the level of progesterone begins to fall; and the level of prolactin rises.
b. and progesterone begin to fall and the level of prolactin rises.
c. falls, the level of progesterone rises, and the level of prolactin falls
d. rises, the level of prolactin falls, and then the level of progesterone rises.

12. When monogamous species of voles in which the male and female both care for the offspring are compared to promiscuous species that do not share parental responsibility

a. there are fewer connections between the MPA and the ventral tegmental area in monogamous males.
b. Fos production in the MPA of monogamous males is lower.
c. the sexual dimorphism of the MPA is less pronounced in the monogamous species.
d. the vasopressin levels in the MPA of monogamous males are lower.

Answers for Self Tests

Lesson I			
1.	b	Obj.	10-1
2.	b	Obj.	10-1
3.	b	Obj.	10-1
4.	d	Obj.	10-2
5.	d	Obj.	10-2
6.	d	Obj.	10-2
7.	c	Obj.	10-3
8.	c	Obj.	10-3
9.	a	Obj.	10-4
10.	b	Obj.	10-4
11.	b	Obj.	10-5
12.	a	Obj.	10-5

Lesson II			
1.	c	Obj.	10-6
2.	c	Obj.	10-6
3.	a	Obj.	10-6
4.	c	Obj.	10-7
5.	b	Obj.	10-7
6.	b	Obj.	10-7
7.	b	Obj.	10-8
8.	d	Obj.	10-8
9.	a	Obj.	10-8
10.	d	Obj.	10-9
11.	a	Obj.	10-9
12.	c	Obj.	10-9

CHAPTER 11
Emotion

Lesson I: Emotions as Response Patterns, Expression and Recognition of Emotions, and Feelings of Emotions

Read the interim summary on pages 334-335 of your text to re-acquaint yourself with the material in this section.

Learning Objective 11-1 Discuss the behavioral, autonomic, and hormonal components of an emotional response and the role of the amygdala in controlling them.

Read pages 325-331 and answer the following questions.

1. List and define the three components of an emotional response in your own words.

    1.

    2.

    3.

2. In general, what is the role of the amygdala in the neural control of emotional responses?

3. List the four major nuclei of the amygdala and their connections. (Study Figure 11.1 in your text.)

Nucleus/Nuclei	Receives information from	Sends information to

4. Underline the nucleus that is most important for expressing emotional responses provoked by aversive stimuli.

5. What behavioral and/or physiological changes occur

    a. to the central nucleus in the presence of aversive stimuli? (Pascoe and Kapp, 1985; Campeau et al., 1991)

b. if the central nucleus is destroyed? (Coover et al., 1992; Davis, 1992b; LeDoux, 1992)

c. from short-term and long-term stimulation of the central nucleus? (Davis, 1992b; Henke, 1982)

6. Review the other regions which receive information from the central nucleus and the responses they control, which are summarized in Figure 11.2 in your text.

7. A classically conditioned response is produced when a(n) _____ stimulus is paired with a stimulus that _____ produces a response and a conditioned emotional response is produced by a(n) _____ stimulus that is paired with a(n) _____-_____ stimulus. The first or _____ response elicited by a painful stimulus is aimed at _____ the stimulus and the second or _____ response involves physiological changes controlled by the autonomic nervous system.

8. a. Define *coping response* in your own words.

    b. What is its effect on a conditioned emotional response?

9. a. Study Figure 11.3 in your text, which diagrams how LeDoux and his colleagues classically conditioned an emotional withdrawal response in rats, and identify the warning stimulus and the emotion-producing stimulus. (reviewed by LeDoux, 1995)

    b. On the day following conditioning, how did the rats respond when they heard the warning tone? What additional response did they make? Be sure to use the term *freezing* in your answer.

    c. What brain region appears to be necessary for a conditioned emotional response to occur? (LeDoux, 1987)

    d. Which response components of conditioned emotional responses are disrupted by lesions of the lateral hypothalamus? the caudal periaqueductal gray matter? (LeDoux et al., 1988)

10. a. Describe the augmented startle response of rats. What do we call the emotion responsible for this phenomenon?

    b. Explain how Davis and his colleagues produced and measured this behavior in rat subjects. (Study Figure 11.4 in your text. Davis, 1992a, 1992b; Davis et al., 1994)

c. Trace the pathway of the augmented startle response by completing the figure below. (Study Figure 11.5 in your text.)

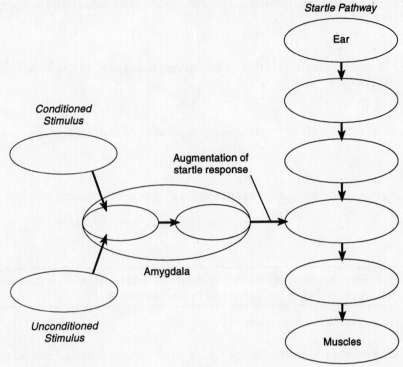

Figure 1

11. Describe how anxiolytic drugs might exert their anxiety-reducing effects by affecting the amygdala.

12. a. How does the neuropeptide CCK affect the brain? Cite research with CCK agonists and antagonists to support your answer.

   b. Summarize the findings of several studies that suggest the amygdala plays a role in some of these effects.

      1. location of CCK-secreting neurons (Ingram et al., 1989; Schiffmann and Vanderhaeghen, 1991)

      2. CCK secretion and fear or anxiety (Pavlasevic et al., 1993; Pratt and Brett, 1995)

      3. CCK agonists and antagonists and signs of anxiety (Frankland et al., 1997)

13. Summarize evidence suggesting the amygdala plays a role in human emotional responses.

   a. Under what circumstances did epilepsy patients report feeling afraid? (White, 1940; Halgren et al., 1978; Gloor et al., 1982)

b. How well do people with lesions of the amygdala learn a conditioned emotional response? (LaBar et al., 1995; Bechara et al., 1995) How does a negative emotion further affect this response? (Angrilli et al., 1996)

c. Compare the memories of normal subjects and people with amygdala damage for emotional story details? (Cahill et al., 1995)

d. Compare the PET scans of people when they recalled neutral films and emotionally arousing films they had seen. (Cahill et al., 1996)

e. And now compare the PET scans of people working on solvable and unsolvable anagrams. (Schneider et al., 1996)

f. How does CCK affect both anxiety and blood flow to the amygdala? (Benkelfat et al., 1995)

Learning Objective 11-2 Discuss the role of the orbitofrontal cortex in the analysis of social situations and the effects of damage to this region, including those produced by psychosurgery.

Read pages 331-334 and answer the following questions.

1. Study Figure 11.6 in your text and describe the location of the orbitofrontal cortex. How did this region get its name?

2. List the direct inputs and the outputs from the orbitofrontal cortex.

Direct Inputs	Outputs
1.	1.
2.	2.
3.	3.
4.	4.
5.	5.

3. a. What kind of information is received through the inputs?

b. What kinds of activities are influenced through the outputs?

4. Summarize early case histories that suggested the role of the orbitofrontal cortex in emotional behavior.

a. What part of Phineas Gage's brain was largely destroyed as a result of his accident and how did his injury affect his behavior? (See Figure 11.7 your text.)

b. To discuss the case of Becky the chimpanzee, begin by describing the experimental task. (Jacobsen et al., 1935)

c. How did Becky react to this task?

d. What kind of lesion did she receive and how did it affect her behavior during subsequent training sessions?

e. How did the medically necessary removal of the frontal lobes of a human appear to affect the patient? (Bricker, 1936)

5. a. What human application did these last two case studies suggest to Egas Moniz? (Fulton, 1949)

b. Following prefrontal lobotomy, what emotional changes were observed in patients?

c. What were some of the unexpected side effects of this surgery?

d. Why was this procedure eventually abandoned and why was surgery on the human brain viewed more skeptically? (Valenstein, 1986)

6. a. Study Figure 11.8 in your text and briefly describe "ice-pick" surgery using a transorbital leucotome.

b. What specific objections did physicians raise to the use of this procedure to perform human brain surgery?

7. a. How well did a patient who had had surgery for the removal of a benign tumor of the orbitofrontal cortex assess hypothetical social situations? (Eslinger and Damasio, 1985)

b. How well did this same patient conduct his personal affairs following surgery?

c. What does this clinical evidence suggest about the role of the orbitofrontal cortex in making judgments and conclusions?

8. a. What are the general functions of the ventral connections of the frontal lobes? the dorsal connections?

b. More specifically, what may be the role of the cingulate gyrus in making appropriate decisions?

c. If the human cingulate gyrus is electrically stimulated, what kinds of feelings occur? (Talairach et al., 1973)

d. If this region is damaged, what syndrome results? (Amyes and Nielsen, 1955)

Read the interim summary on pages 341-342 of your text to re-acquaint yourself with the material in this section.

Learning Objective 11-3 Discuss cross-cultural studies on the expression and comprehension of emotions.

Read pages 335-337 and answer the following questions.

1.   Why is the expression and recognition of emotions a beneficial social behavior?

2.   a. State Darwin's hypothesis concerning the origin of human facial expression of emotion. (Darwin, 1872/1965)

b. What evidence did he obtain to support his conclusion?

3.   Describe modern research by (Ekman and Friesen, 1971; Ekman, 1980) that tends to confirm Darwin's hypothesis.

a. Who were the subjects?

b. How easily did they recognize the facial expression of westerners?

c. How easily did westerners recognize the facial expressions of a man from this tribe shown in Figure 11.9 in your text?

4.   Explain the rationale for and the results of research comparing the facial expressions of young blind and sighted children, but not of blind and sighted adults. (Woodworth and Schlosberg, 1954; Izard, 1971)

5.   What do the consistent patterns of the facial expression of emotion supported by cross-cultural studies and studies with the blind suggest about their basis?

6.   a. Define display rules in your own words.

b. According to Ekman and Friesen (1975), how do display rules affect emotional expression?

7.   a. During the showing of a coming-of-age rite how did American and Japanese students reach when they were alone? with others?

b. What do these findings suggest about the influence of culture on the display of emotion?

Learning Objective 11-4 Discuss the neural control of the recognition of emotional expression in normal people and people with brain damage.

Read pages 337-339 and answer the following questions.

1. a. What were some of the situations in which Kraut and Johnston (1979) observed the emotional expression of subjects?

    b. When did the subjects show the greatest reaction?

    c. What does this study suggest about emotional expression and communication?

2. If the right hemisphere plays a more important role in the comprehension of emotion, why is it advantageous to receive emotional stimuli in the left ear or left visual field?

3. Briefly review some of the tasks for which the right and the left hemisphere is better suited. (reviewed by Bryden and Ley, 1983)

4. a. Which recognition tasks did patients with right-hemisphere lesions find difficult? did not find difficult? (Blonder et al., 1991; Bowers et al., 1991)

    b. What do these responses suggest about the role of the right-hemisphere in the comprehension of emotions conveyed by real or imagined facial expression or hand gestures?

5. a. Describe the three situations George et al. (1996) used to test comprehension of emotion.

    b. As recorded by PET scans, when did the right and left hemispheres of the brain show increased activity? (See Figure 11.10 in your text.)

6. a. What kind of brain damage had subjects in a study by Heilman et al., (1975) received?

    b. Describe the experimental task and the accuracy of the subjects' response.

    c. How accurately did a man with pure word deafness identify the emotional content of speech? (Heilman et al., 1983)

    d. What do these studies suggest about the components of comprehension of emotion?

7. What are some of the combinations of recognition deficits for faces and facial expressions of emotion experienced by people with damage to visual association cortex? (Bowers and Heilman, 1981; Humphreys et al., 1993)

8. a. What characteristic of the face do neurons in a monkey's superior temporal sulcus appear to recognize? (Study Figure 11.11 in your text. Perrett et al., 1992)

    b. Why is gaze important in the recognition of emotions?

    c. How do lesions of the superior temporal sulcus affect monkeys' gaze recognition? (Campbell et al., 1990; Heywood and Cowey, 1992)

    d. Which neural connections may play a role in this recognition? (Harries and Perrett, 1991)

9. Summarize evidence that the amygdala may also play a role in the recognition of facial expressions of emotion. (For example, Scott et al., 1997)

Learning Objective 11-5 Discuss the neural control of emotional expression in normal people and people with brain damage.

Read pages 339-341 and answer the following questions.

1. Why is it physiologically impossible to voluntarily produce a genuine smile? (Duchenne, 1862/1990)

2. a. What kind of voluntary and involuntary facial movements can patients with volitional facial paresis make? patients with emotional facial paresis. (Representative facial movements are shown in Figure 11.12 in your text. Hopf et al., 1992)

    b. What kind of brain damage causes each of these syndromes?

    c. What do these syndromes indicate about the movement of facial muscles and the genuine expression of emotion?

3. The _____ hemisphere appears to be specialized for both the _____ of emotion and the _____ of emotion.

4. a. How are chimerical faces, shown in Figure 11.13, created?

b. Why is the left half of the face more expressive?

c. What more natural observations confirmed this? (Moscovitch and Olds, 1982)

d. When the chimerical faces technique was used with rhesus monkeys, what did the analysis of the videotapes indicate about hemispheric specialization

  1. of emotional expression? (Study Figure 11.14 in your text.)

  2. in the evolution of emotional expression? (Hauser, 1993)

5. Compare the ability of patients with Wernicke's aphasia and patients with right hemisphere lesions to express emotions using tone of voice.

6. a. When and how does a physician perform a Wada test?

   b. What did Ross and his colleagues observe during a Wada test? (Ross et al., 1994)

Read the interim summary on pages 344-345 in your text to re-acquaint yourself with the material in this section.

Learning Objective 11-6 Discuss the James-Lange theory of feelings of emotion and evaluate relevant research.

Read pages 342-344 and answer the following questions.

1. Outline the James-Lange theory in your own words. (James, 1884; Lange, 1887)

2. Carefully study Figure 11.15 in your text that diagrams the James-Lange theory. Check your understanding by describing the process.

3. According to the theory, in what order do these two events produce feelings of emotion?

   "I'm more nervous than I thought I was." / Her stomach felt queasy as she waited to be interviewed for a second time for the job.

4. In what way does the James-Lange theory appear to contradict personal experience?

5. a. Explain two of Cannon's objections to the James-Lange theory. (Cannon, 1927)

    b. Now refute his objections.

6. Why is the theory difficult to verify?

7. a. Describe the subjects interviewed by Hohman (1966).

    b. Explain how he tested the James-Lange theory by studying these subjects.

    c. What was the relationship between the level of injury and the intensity of feelings of emotion? Explain.

8. a. Why did Ekman and his colleagues ask subjects to move particular facial muscles, but gave them no further information? (Ekman et al., 1983; Levenson et al., 1990)

    b. What, according to the physiological monitoring, happened to the subjects while they made these movements?

    c. Suggest two explanations for the results you have just described.

9. Study Figure 11.16 in your text to see the facial expressions posed by adults in front of infants and the infants' responses. What do their responses suggest about the tendency to imitate? (Field et al., 1982)

Lesson I Self Test

1. If an animal learns how to avoid, escape from, or minimize an aversive stimulus, it has learned a(n)

    a. conditioned emotional response.
    b. coping response.
    c. offensive emotional response.
    d. defensive emotional response.

2. If the central nucleus of the amygdala is destroyed,

    a. autonomic and behavioral components of conditioned emotional responses are disinhibited.
    b. conditioned emotional responses to visual and olfactory, but not auditory, stimuli are disrupted.
    c. positive and negative feelings associated with conditioned emotional responses are disrupted.
    d. conditioned emotional responses cannot be learned.

3. An augmented startle response occurs through a connection between the central nucleus of the amygdala and the

   a. ventral nucleus of the lateral lemniscus.
   b. caudate nucleus.
   c. nucleus reticularis pontis caudalis.
   d. dorsolateral nucleus of the thalamus.

4. People who have suffered damage to the orbitofrontal cortex

   a. suffer from compulsive behaviors.
   b. do not exhibit normal timidity in strange situations.
   c. respond appropriately to hypothetical social situations, but not when these situations apply to them.
   d. show a tendency to express emotional feelings using gestures and facial expressions rather than verbally.

5. What may be the function of the orbitofrontal cortex?

   a. to control voluntary activity
   b. to organize hormonal responses to emotional stimuli
   c. to make judgments and conclusions
   d. to translate judgments into appropriate feelings and behaviors

6. Accurate identification of the facial expressions of Westerners by members of an isolated New Guinea tribe tends to confirm Darwin's hypothesis that emotional expressions

   a. are innate, unlearned responses.
   b. consists of four responses: fear, anger, sorrow, and surprise.
   c. are immune to the effects of socialization.
   d. are identical, whether posed or spontaneous.

7. The _____ hemisphere is involved in _____ emotions and the _____ hemisphere is involved in _____ emotions.

   a. left; genuine; right; posed
   b. right; negative; left; positive
   c. left; fleeting; right; longer lasting

   d. right; verbal expression of; left; nonverbal expression of

8. Following right hemisphere damage, patients

   a. have difficulty determining the correct emotion conveyed in a situation.
   b. can still imagine and describe mental images of emotions.
   c. have difficulty expressing emotion with the face and voice.
   d. can answer questions about emotional and nonemotional situations.

9. People with emotional facial paresis cannot _____ facial expressions of emotion.

   a. imagine
   b. mimic
   c. spontaneously produce
   d. distinguish between different

10 Research results using the chimerical faces technique suggest that the _____ half of the _____ is _____ expressive.

   a. left; brain; more
   b. left; face; less
   c. right; brain; more
   d. right; face; more

11. According to the James-Lange theory, emotional feelings

   a. result from sensory feedback from the responses of emotion-producing situations.
   b. are a direct response to emotion-producing situations.
   c. are a product of both sensory feedback and acquired social behavior.
   d. result in emotional behavior.

12. The results of research on patients with spinal cord injuries suggests that the intensity of their emotional states is related to the

   a. frequency of social contact.
   b. level of injury to the spinal cord.
   c. length of time following injury
   d. perception of self-worth.

Lesson II: Aggressive Behavior

Read the interim summary on pages 353-354 in your text to re-acquaint yourself with the material in this section.

Learning Objective 11-7 Discuss the nature, functions, and neural control of aggressive behavior.

Read pages 345-348 and answer the following questions.

1. Complete these statements.

    a. Aggressive behaviors are species-typical; that is

    b. Many aggressive behaviors are related to

2. List the three basic forms of aggressive behavior and give an example of each kind of behavior.

    1.

    2.

    3.

3. a. Define *predation* in your own words.

    b. Compare the level of arousal and activity of the autonomic nervous system of animals engaged in offensive or defensive behaviors and predatory behaviors.

4. a. The neural control of aggressive behavior is _____.

    b. In general, where are the neural circuits for the particular movements an attacking or defending animal makes located?

    c. Which structures appear to control these circuits?

    d. What controls the activity of the limbic system?

5. a. What task did Roberts and Kiess (1964) teach laboratory cats?

    b. What was the only situation in which these cats would seek out a rat?

    c. Describe the behavior of a hungry cat when brain stimulation was turned on.

    d. What do these results suggest about the nature of feeding and predatory attack and the neural control of eating and attack?

6. Brain stimulation that elicits _____ attack appears to be aversive, and brain stimulation that elicits _____ attack appears to be reinforcing. (Panksepp, 1971)

7. a. In general, how did electrical or chemical stimulation of the periaqueductal gray matter (PAG) affect aggressive behavior?

    b. What other brain structures may play a role in these behaviors?

8. a. More specifically, how is aggressive behavior affected

    1. if, while the dorsal PAG is being stimulated, the medial hypothalamus is also stimulated? (Schubert et al., 1996)

    2. if AP-7 is infused into the dorsal PAG?

    b. How did the researchers establish that there was a connection between the dorsal PAG and the medial hypothalamus?

    c. Study Figure 11.17 in your text and note the regions of the amygdala and the hypothalamus that are involved in defensive rage and predation.

9. Why are serotonergic drugs sometimes used to treat violent behavior in humans?

10. If serotonergic axons in the forebrain are destroyed, how is aggressive attack affected? (Vergnes et al., 1988)

11. a. How did researchers assess the level of serotonergic activity in the brains of monkeys living in a free-ranging colony? (Mehlman et al., 1995; Higley et al., 1996a, 1996b.)

    b. To be sure that you understand what happens when serotonin (5-HT) is released in the brain, explain the relationship between the level of 5-HIAA in cerebrospinal fluid and the level of 5-HT.

    c. What kinds of activities did young male monkeys with low levels of 5-HIAA engage in?

    d. What was their survival rate? (Study Figure 11.18 in your text.)

    e. What do these results suggest about the role of serotonin in aggressive behavior?

12. a. How did dominance patterns in a monkey colony change when the dominant males were removed and lower ranked males received either a serotonin agonist or antagonist? (Raleigh et al., 1991)

    b. Carefully explain why these results suggest that dominance and aggression are not the same thing.

13. a. In humans, what is the relationship between an individual's 5-HIAA level and

    1. aggressiveness and antisocial tendencies? (Brown et al., 1979; 1982)

2. behavioral problems among other close relatives? (Coccaro et al., 1994)

b. Briefly summarize the research to determine the role of MAO type A in aggressive behavior. (Brunner et al., 1993)

Learning Objective 11-8   Discuss the hormonal control of intermale aggression, interfemale aggression, maternal aggression, and infanticide.

Read pages 348-351 and answer the following questions.

1. Explain what the emergence of intermale aggression at puberty suggests about the control mechanisms of aggressive behaviors. Cite research to support your answer. (Beeman, 1947)

2. a. Briefly review the organizational effect of early androgenization that influences both intermale aggression and male sexual behavior.(Study Figure 11.19 in your text.)

b. Early androgenization _____ neural circuits.

3. If a treatment such as prenatal stress interferes with prenatal masculinization, what two behavioral changes are observed in male offspring at adulthood? (Kinsley and Svare, 1986)

4. a. How did castrated male rats respond to testosterone implants in the medial preoptic area? (Bean and Conner, 1978)

b. What does this research suggest about one of the roles of the MPA? What other role have you previously studied?

5. a. What technique did Bean (1982) find abolished intermale aggression in mice? What technique had no effect? (Study Figure 11.20 in your text.)

b. What is the probable stimulus for intermale aggression and what brain region may play a role?

c. What other type of aggressive behavior appear to be controlled by pheromones? (Dixon and Mackintosh, 1971; Dixon, 1973)

6. Discuss evidence that interfemale aggression is also dependent on testosterone.

a. How did ovariectomized female rats respond to injections of testosterone? (Study Figure 11.21 in your text. Van de Poll et al., 1988)

b. Study Figure 11.22 in your text and be sure you understand the difference between 0M, 1M, and 2M females.

c. Which of these females have the highest levels of prenatal testosterone and later exhibit the most interfemale aggression? (Vom Saal and Bronson, 1980)

7. At what time in the estrous cycle are some primates most likely to engage in aggressive behavior with males? (Carpenter, 1942; Saayman, 1971) with females? (Sassenrath et al., 1973; Mallow, 1979)

8. Summarize some of the findings of a literature review on premenstrual syndrome (PMS). (Floody, 1983)

   a. At what times in the menstrual cycle did observed aggressiveness increase and decrease?

   b. How widespread do mood shifts and actual aggressiveness appear to be?

   c. Which women were more likely to experience them? (Persky, 1974)

9. a. Compare the latency to attack an intruder by a lactating female with that of two strange males encountering each other. (Svare, 1983)

   b. When do pregnant mice begin to become aggressive? Why?(Mann et al., 1984)

   c. What change in aggressiveness occurs immediately after giving birth? Why? (Ghiraldi and Svare, 1989; Svare, 1989)

   d. How can this period of docility be interrupted? restored?

   e. What two stimuli provided by offspring activate aggressive behavior?

   f. How can one of these stimuli be eliminated? (Svare and Gandelman, 1976; Gandelman and Simon, 1980; Svare et al., 1982)

   g. Briefly summarize how prenatal exposure to androgens appears to affect maternal aggression. (Vom Saal and Bronson, 1980a; Kinsley et al. 1986)

10. a. Why, according to Hrdy (1977), may a male langur kill the infants of another male?

   b. Briefly summarize how the length of time between copulations appears to regulate the tendency of male mice to kill infants of other males. (Vom Saal, 1985)

   c. How did Perrigo and colleagues (Perrigo et al., 1990) determine what the units of time that regulate infanticide are?

11. Discuss two reasons and supporting research on why females commit infanticide. (Calhoun, 1962; Gandelman and Simon, 1978)

Learning Objective 11-9   Discuss the effects of androgens on human aggressive behavior.

Read pages 351-353 and answer the following questions.

1. Explain why the study of human male aggression must consider the effects of both socialization and androgenization.

2. a. Briefly describe the effects of castration of convicted male sex offenders. (Hawke, 1951; Sturup, 1961; Laschet, 1973)

    b. In what way were these studies flawed?

3. What alternative treatment for human sexual aggression is preferable and how effective is it? (Walker and Meyer, 1981; Zumpe et al., 1991)

4. a. What was the conclusion of a literature review concerning the relationship between men's' testosterone levels and the level of aggression? (Archer, 1994)

    b. Compare the testosterone levels of male prisoners and female prisoners who engaged in unprovoked violence. (Morris, 1990; Dabbs et al., 1987; Dabbs et al., 1988)

5. Carefully explain why we cannot conclude that high testosterone levels cause increased aggression. Cite research to support your explanation. (Mazur and Lamb, 1980; Elias, 1981; McCaul et all, 1992; Jeffcoate et al., 1986)

6. a. Now explain the ethical concerns that prevent the study of the effects of testosterone by giving experimental subjects androgen supplements.

    b. What is the only kind of experimental evidence on this subject available? Cite an example.

7. a. How did anabolic steroids affect the aggressive behavior of male weight lifters? (Yates et al., 1992)

    b. Why is it incorrect to conclude that steroids are responsible for increased aggressiveness?

8. a. How and when does alcohol intake affect intermale aggression in dominant male squirrel monkeys? subordinate monkeys? (Study Figure 11.23 in your text. Winslow and Miczek, 1985, 1988)

    b. What do these effects suggest about factors that influence aggressive behavior?

c. What did tests during the nonmating season confirm? (Winslow et al., 1988)

Lesson II Self Test

1. Predatory attack

   a. is usually accompanied by a strong display of rage.
   b. and eating are organized by different neural mechanisms.
   c. elicited by electrical stimulation appears aversive to laboratory animals.
   d. on rats by laboratory cats is a spontaneous behavior.

2. If AP-7 is infused into the periaqueductal gray matter, the effects of medial hypothalamic stimulation are

   a. increased.
   b. blocked.
   c. delayed.
   d. inconsistent.

3. Increased activity of serotonergic synapses _____ aggression.

   a. increases
   b. inhibits
   c. initiates
   d. has no effect on

4. Monkeys in a free-ranging colony with the lowest levels of a metabolite of serotonin

   a. showed increased risk-taking behavior.
   b. had the longest survival rates.
   c. usually became the dominant monkeys.
   d. had the highest levels of social competency.

5. As adults, the male offspring of prenatally stressed females exhibited

   a. less intermale aggression.
   b. more intermale aggression.
   c. less affect during predatory attack.
   d. a marked aversion to all forms of aggressive behavior.

6. Females who were next to a male fetus in the uterus _____ than females located between two females.

   a. had significantly higher levels of testosterone in their blood

   b. were less likely to display female sexual behavior.
   c. were more likely to attack a male
   d. exhibited less maternal behavior

7. Females of some primate species are more likely to engage in fights

   a. during interruptions of menstruation caused by events such as pregnancy or low food supply.
   b. just before and just after menstruation.
   c. around the time of ovulation and just before menstruation.
   d. following ovulation if pregnancy does not occur.

8. Immediately after giving birth, female mice

   a. exhibit heightened irritability and aggressiveness.
   b. are more hostile toward rival females than males.
   c. will attack only males.
   d. are docile for approximately 48 hours.

9. Infanticide by males is regulated by

   a. pheromones.
   b. proximity of rival males.
   c. copulation.
   d. blood levels of testosterone.

10. Studies of the effects of castration on human aggression

    a. are the only way to determine the effects of androgens on aggression.
    b. usually do not measure aggressive behavior directly.
    c. do not need to include control groups.
    d. usually do not correct for the effects of age at the time of castration on aggressive behavior.

11. Studies to correlate blood levels of testosterone and aggression

    a. indicate that high levels of testosterone cause increased aggressive behavior.
    b. are more reliable for male subjects than female subjects.

c. must take into account a person's
    environment in making conclusions.
d. cannot be undertaken because of ethical
    concerns.

12. Alcohol increases intermale aggression

a. among all male squirrel monkeys.

b. among dominant male squirrel monkeys but
    only during the mating season.
c. among subordinate male squirrel monkeys.
d. among all male squirrel monkeys until
    successful copulation occurs.
d. during retaliatory attack.

Answers for Self Tests

Lesson I		Lesson II	
1. b Obj. 11-1		1. b Obj. 11-7	
2. d Obj. 11-1		2. b Obj. 11-7	
3. c Obj. 11-1		3. b Obj. 11-7	
4. c Obj. 11-2		4. a Obj. 11-7	
5. d Obj. 11-2		5. a Obj. 11-8	
6. a Obj. 11-3		6. a Obj. 11-8	
7. b Obj. 11-4		7. c Obj. 11-8	
8. c Obj. 11-4		8. d Obj. 11-8	
9. c Obj. 11-5		9. c Obj. 11-8	
10. c Obj. 11-5		10. b Obj. 11-9	
11. a Obj. 11-6		11. c Obj. 11-9	
12. b Obj. 11-6		12. b Obj. 11-9	

CHAPTER 12
Ingestive Behavior: Drinking

Lesson I: Fluid Balance, Drinking, and Salt Appetite

Read the interim summary on page 361 of your text to re-acquaint yourself with the material in this section.

Learning Objective 12-1   Explain the characteristics of a regulatory mechanism.

Read pages 356-357 and answer the following questions.

1. Define *homeostasis* and *ingestive behavior* in your own words and explain their importance.

2. What is the function of a regulatory mechanism?

3. List and explain the functions of the four essential features of a regulatory mechanism. (See Figure 12.1 in your text.)

    1.   3.

    2.   4.

4. Using the example of the room thermostat, explain the process of negative feedback.

5. a. What role do ingestive behaviors play in homeostasis?

    b. Study Figure 12.2 in your text and explain the relation of the satiety mechanism to the correctional mechanism in the control of drinking.

Learning Objective 12-2   Describe the fluid compartments of the body and describe how the kidneys control the excretion of water and sodium.

Read pages 357-361 and answer the following questions.

1. There are _____ major fluid compartments in the body--one for _____ fluid and _____ for

_____ fluid. The intracellular fluid is the fluid portion of the _____ of cells and contains approximately _____ percent of the body's water. The extracellular fluid includes the _____ fluid or blood plasma, the _____ fluid of the brain, and the _____ fluid between our cells. (Study Figure 12.3 in your text.)

2. Use the terms *isotonic, hypertonic,* and *hypotonic* to explain why the concentration of the interstitial fluid must remain constant. (Study Figure 12.4 in your text.)

3. a. Explain why the volume of blood plasma must be closely regulated by describing the consequences of hypovolemia.

   b. What limited correctional mechanism does the body use when blood volume is too low?

4. Explain why the volume of the interstitial fluid need not be regulated so closely, but it's tonicity must be.

5. a. Why are intracellular fluid and blood volume monitored by two different sets of receptors?

   b. What correctional mechanisms are controlled by these receptors?

6. What is the general function of the kidneys and the specific function of the nephrons and the ureter? (See Figure 12.5 in your text.)

7. a. Which steroid hormone regulates sodium excretion?
   b. Where is it produced, and how is it released into the blood stream?

   c. What is the effect of high levels of this hormone on the kidneys? low levels? (See Figure 12.6 in your text.)

8. a. Which peptide hormone regulates water excretion by the kidneys?

b. Where is it produced, stored, and released?

c. What is the effect of high levels of this hormone on the kidneys? low levels?

d. If this hormone is lacking, what disease results?

Read the interim summary on page 367 of your text to re-acquaint yourself with the material in this section.

Learning Objective 12-3 Explain the control of osmometric thirst.

Read pages 361-364 and answer the following questions.

1. _____ thirst occurs when the tonicity of the interstitial fluid _____.

2. Study Figures 12.7 and 12.8 in your text and explain how the size and firing rate of an osmoreceptor changes as the surrounding interstitial fluid becomes more concentrated. Be sure that you understand the movement of water in osmosis. (Verney, 1947)

3. Now study Figure 12.9 in your text and explain how our bodies loose water from all three fluid compartments through evaporation, and the accompanying changes that occur.

4. Describe what happens, step by step, to the fluid compartments of the body when we eat a salty meal.

5. a. Why did Fitzsimons (1972) begin his study of the stimulus for osmometric thirst by removing the kidneys of laboratory rats?

   b. One group of subject rats was injected with a substance that could enter cells and another group was injected with a substance that could not enter cells. Which group drank excessively and why?

   c. What, then, is the stimulus for osmometric thirst?

   d. When Fitzsimons injected subject rats with urea, how was drinking affected and why?

   e. What do these results suggest about the general location of osmoreceptors?

6. a. Which more specific locations have been suggested by research by

      1. Andersson (1953)?

      2. Peck and Blass, (1975)?

3. Blass and Epstein (1971)?

b. What differing results have been obtained by

  1. bilateral injections of hypertonic and hypotonic solutions in to the preoptic area? (Andrews et al., 1992)

  2. destruction of the lateral preoptic area? (for example, Said et al., 1996)

7. a. Where do most researchers believe the osmoreceptors are located?

  b. Name several circumventricular organs. Describe their blood supply and their location and its significance.

Learning Objective 12-4 Explain the control of volumetric thirst.

Read pages 364-367 and answer the following questions.

1. _____ thirst occurs when the volume of blood plasma _____.

2. a. Why does evaporation produce both osmometric and volumetric thirst?

  b. Identify three conditions that cause volumetric thirst.

3. a. Now describe a procedure for initiating volumetric thirst in experimental animals. What kind of substance is injected into the animal?

  b. What property of colloid draws extracellular fluid and sodium out of tissue gradually creating a vacant space?

  c. What fluid begins to move into this vacant space and why?

  d. In response to changing blood volume, what changes occur in the posterior pituitary gland and in the kidneys?

  e. Throughout this procedure, what changes, if any, took place in the cells?

4. a. After injecting subject rats with polyethylene glycol, a colloid, what did Fitzsimons (1961) do?

  b. Describe how rats drank after the draining procedure and the next day.

  c. The lost of sodium induced a _____ _____ in the rats.

5. Fill in the blanks in Figure 1 below.

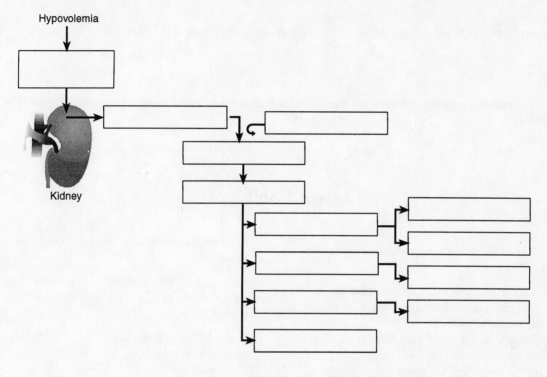

Figure 1

6. a. In addition to the kidneys, where is a second set of receptors for volumetric thirst found?

    b. Briefly explain how atrial baroreceptors in the heart detect changes in blood volume. Cite research to support your answer. (Fitzsimons and Moore-Gillon, 1980; Moore-Gillon and Fitzsimons, 1982; Quillen et al., 1990)

7. a. Why do eating and drinking usually occur together?

    b. How did a change in diet gradually affect the drinking behavior of rats? (Fitzsimons and Le Magnen, 1969)

8. a. Explain how eating a normally sized meal soon leads to thirst and drinking. Cite research to support your answer. (Nose et al., 1986; Rowland, 1995; Kraly and Corneilson, 1990)

    b. What may be the role of histamine in food-related drinking? (For example, Kraly and June, 1982)

    c. How is drinking affected by cutting the nerves to the kidney? Why? (Kraly et al., 1995)

9. a. What is the primary stimulus for a salt appetite?

b. What two other signals may also be involved?

c. Why is the role of angiotensin still uncertain? (For example, Weisinger et al., 1996)

d. What evidence suggests that the atrial baroreceptors function independently of angiotensin? (Thornton et al., 1994)

Lesson I Self Test

1. The role of the detector in a regulatory mechanism is to

   a. monitor the value of the system variable.
   b. recognize stimuli that might change the value of the system variable.
   c. establish the value of the set point in response to environmental conditions.
   d. restore the system variable to the set point.

2. An essential feature of all regulatory mechanisms is

   a. a satiety mechanism.
   b. negative feedback.
   c. continuous feedback.
   d. a neural control mechanism.

3. Satiety mechanisms

   a. monitor system variables.
   b. replenish depleted stores of food, fluid, or nutrients.
   c. are a second correctional mechanism if detectors fail to work properly.
   d. monitor the activity of correctional mechanisms.

4. Approximately two-thirds of the body's water is contained in the _____ fluid.

   a. intracellular
   b. extracellular
   c. intravascular
   d. interstitial

5. If the fluid inside a cell is isotonic to the interstitial fluid

   a. their volumes are equal.
   b. water will not tend to move in or out.
   c. water will diffuse out of the cells through osmosis.

   d. osmotic pressure may cause the membrane to rupture.

6. The _____ of the interstitial fluid must be closely regulated, but its _____ usually remains within normal limits.

   a. production; excretion
   b. excretion; volume
   c. tonicity; volume
   d. volume; tonicity

7. If we drink more water than we need, _____ secretion of _____ causes the kidneys to excrete more _____.

   a. decreased; vasopressin; water
   b. deceased; aldosterone; water
   c. increased; vasopressin; sodium
   d. increased; aldosterone; sodium

8. Osmometric thirst occurs when the

   a. tonicity of the interstitial fluid increases.
   b. volume of the intravascular fluid decreases.
   c. blood flow to the kidneys decreases.
   d. blood level of renin increases.

9. The *stimulus* for osmometric thirst is

   a. renin secretion
   b. a decrease in urine production.
   c. hypovolemia.
   d. cell dehydration.

10. The easiest way to produce volumetric thirst in an experimental animal is to give an injection of

   a. a colloid.
   b. a saline solution.
   c. saralasin.
   d. losartan.

11. A behavioral effect of angiotensin II is

    a. aversion to a high protein diet.
    b. production of a salt appetite.
    c. near cessation of drinking.
    d. an absence of food-related drinking.

12. The detectors for volumetric thirst are located in the

    a. posterior pituitary gland and the hypothalamus.
    b. kidneys and the bladder.
    c. kidneys, heart, and large blood vessels.
    d. stomach and large intestine.

Lesson II: Brain Mechanisms of Thirst and Salt Appetite and Mechanisms of Satiety

Read the interim summary on page 372 in your text to re-acquaint yourself with the material in this section.

> *Learning Objective 12-5* Describe the neural control of thirst and a salt appetite.

Read pages 367-372 and answer the following questions.

1. a. In which part of the brain is the nucleus of the solitary tract found?

   b. Which parts of the body send information to it and where does it send its efferent axons? (Johnson and Edwards, 1990)

2. Briefly summarize research that the region of the AV3V may play a role in fluid regulation in rats (Thornton et al., 1984) and humans (McIver et al., 1991).

3. a. What property of angiotensin suggests that it may produce thirst through one of the circumventricular organs?

   b. Which circumventricular organ appears to be the site of action of blood angiotensin? Cite research to support your answer (For example, Smith et al., 1995. See Figure 12.11 in your text.)

4. a. What is the primary role of the subfornical organ (SFO)?

   b. The outputs of the SFO can be divided into three categories. List them and their projections.

      1.   Projections

      2.   Projections

      3.   Projections

5. Review the five major effects of angiotensin.

6. a. How do lesions of the ventral stalk of the SFO affect drinking induced by injections of AII into a vein? (Lind et al., 1984) into the third ventricle?

   b. What do these results suggest about the location of other angiotensin receptors? (See Figure 12.11.)

7. Outline Lind and Johnson's (1982) explanation for the presence of angiotensin receptors in the median preoptic nucleus. Cite supporting evidence for this role. (Johnson and Cunningham, 1987; Tanaka and Nomura, 1993)

8. Outline the explanation suggested by Thrasher and his colleagues (Thrasher, 1989) of the role of the region in front of the third ventricle in osmometric and volumetric thirst. (Study Figure 12.12 in your text.)

9. a. How do neurotoxic lesions of the median preoptic nucleus affect drinking caused by angiotensin and hypertonic saline? (See Figure 12.13 in your text. Cunningham et al., 1992)

   b. What do these results suggest about other controls of drinking?

10. Summarize research, using various techniques, to learn more about the role of the following structures in drinking.

    a. lesions of the lateral hypothalamus (Teitelbaum and Epstein, 1962)

    b. electrical stimulation of the rostral zona incerta (Huang and Mogenson, 1972) Refer to Figure 10.19 in your text for its location.

    c. lesions of the zona incerta (Walsh and Grossman, 1978)

    d. recordings of single neurons in the zona incerta (Mok and Mogenson, 1986)

    e. injections of angiotensin into a cerebral ventricle (Czech and Stein, 1992)

11. Outline the neural control of salt appetite by describing evidence

    a. to identify some brain structures involved in the behavioral effects of aldosterone on salt appetite. (Coirini et al., 1985; Schulkin et al., 1989. See Figure 12.14 in your text.)

    b. that aldosterone is not the only stimulus for salt appetite. Be sure to mention furosemide in your answer.

    c. on the role of both anterior circumventricular organs in salt appetite. (Weisinger et al., 1990; Fitts et al., 1990)

    d. for the role of the zona incerta. (Grossman and Grossman, 1978; Gentil et al., 1971)

Read the interim summary on page 374-375 of your text to re-acquaint yourself with the material in this section.

> *Learning Objective 12-6* Describe the satiety mechanism for drinking caused by drinking and ingestion of sodium chloride.

Read pages 372-374 and answer the following questions.

1. Summarize evidence that receptors in the mouth and throat play a role in satiety for drinking.

    a. What is the delay between a thirsty dog drinking its fill and the water level of the blood plasma being replenished? (Adolph, 1939; Ramsay et al., 1977)

    b. In research by Miller et al., (1957), some rats were permitted to drink water, but how did others receive water?

    c. Which rats later drank more water? (See Figure 12.15 in your text.)

    d. What do these results suggest about the function of receptors in the mouth and throat?

2. a. How have esophageal fistulas been used to study the duration of satiety produced by these receptors?

    b. What were the results of studies using this procedure?

3. a. Why did researchers place a noose around the pylorus of subject rats as illustrated in Figure 12.16 in your text? (Hall, 1973; Hall and Blass, 1977)

    b. When the noose was tightened, how was drinking and satiety receptors in the stomach, liver and duodenum affected?

    c. What do these results suggest about the importance of particular receptors which signal satiety?

4. a. Which part of the small intestine receives food and water from the stomach?

    b. Study Figure 12.17 in your text and then explain how water and nutrients move from the small intestine to the liver. Be sure to mention the hepatic portal vein in your answer.

5. What is known about the role of receptors in the duodenum in satiety?

6. a. How did infusions of water into the hepatic portal vein affect

    1. osmometric drinking initiated by injections of hypertonic saline? (Kozlowski and Drzewiecki, 1973)

    2. drinking in water deprived rats? (Kobashi and Adachi, 1992)

    b. How did infusions of isotonic or hypertonic saline affect these same rats?

    c. What do these results suggest about role of the osmoreceptors in the liver?

    d. What nerve may carry inhibitory signals for thirst to the brain? (Kobashi and Adachi, 1993)

7. Finally let's look at satiety receptors for salt appetite. Which receptors do not appear to be important in satiety for salt appetite? (Mook, 1969; Wolf et al., 1984)

8. a. How did injections of a hypertonic sodium chloride solution into the hepatic portal vein affect the amount of a salt solution an animal would drink? (Tordoff et al., 1987)

    b. What do the results suggest about the role of receptors in the liver?

9. a. Where is atrial natriuretic peptide (ANP) secreted?

    b. Begin a list of its functions.

    c. Under what circumstances is it secreted?

    d. Where are ANP receptors found?

    e. What structure plays a role in the inhibitory effect of ANP on drinking? (Quirion et al., 1984; Nermo-Lindquist et al., 1990; Ehrlich and Fitts, 1990)

Lesson II Self Test

1. The region around the anteroventral third ventricle of the brain including the OVLT

   a. monitors evaporation rate through skin temperature receptors.
   b. contains osmoreceptors that stimulate thirst and vasopressin secretion.
   c. initiates hypovolemia.
   d. is the weakest portion of the blood-brain barrier.

2. Angiotensin is _____ which _____ the blood brain barrier.

   a. both a peptide and a hormone; does not cross
   b. a peptide; crosses
   c. both a hormone and a transmitter substance; crosses
   d. an enzyme; does not cross

3. Angiotensin

   a. blocks vasopressin secretion.
   b. lowers blood pressure.
   c. stimulates aldosterone secretion.
   d. increases urine output.

4. The subfornical organ

   a. does not appear to have any angiotensin receptors.
   b. secretes saralasin which blocks angiotensin.
   c. is the site of action of angiotensin.
   d. breaks down angiotensin.

5. The effect of the subfornical organ on drinking occur through its _____ outputs

   a. autonomic
   b. behavioral
   c. endocrine
   d. metabolic

6. The angiotensin receptors in the median preoptic nucleus

   a. are stimulated by angiotensin that crosses the blood-brain barrier.
   b. detect angiotensin secreted as a transmitter substance by terminal buttons there.
   c. are an example of the redundant systems that have evolved in the brain.
   d. control the secretion of vasopressin

7. Lesions of the zona incerta

   a. have no effect on the stimuli for volumetric thirst.
   b. stimulate salt intake.
   c. abolish drinking in response to injections of the colloid polyethylene glycol.
   d. produce a profound deficit in osmometric drinking.

8. Where does aldosterone appear to exert its behavioral effects?

   a. medial nucleus of the amygdala
   b. lateral hypothalamus
   c. zona incerta
   d. median preoptic nucleus

9. Rats who received preloads of water in the mouth drank _____ than rats who received the preloads in the stomach.

   a. more
   b. less
   c. the same amount of water
   d. the same amount of water, but more slowly

10. Research that used an esophageal fistula to prevent water from reaching the stomach suggest that satiety receptors in the mouth and throat

    a. play a greater role that first believed.
    b. have a short-lived effect.
    c. cannot signal satiety if connections with stomach receptors are blocked.
    d. and stomach are of equal importance.

11. When the pylorus of an experimental animal is constricted by a noose stomach contents

    a. back up into the esophagus and throat
    b. fall to the ground.
    c. continue to diffuse into the blood through the capillaries.
    d. cannot leave the stomach.

12. Atrial natriuretic peptide stimulates the

    a. excretion of sodium.
    b. secretion of renin.
    c. retention of water.
    d. elevation of blood volume.

Answers for Self Tests

Lesson I		
1.	a	Obj. 12-1
2.	b	Obj. 12-2
3.	d	Obj. 12-2
4.	a	Obj. 12-2
5.	b	Obj. 12-2
6.	c	Obj. 12-2
7.	a	Obj. 12-2
8.	a	Obj. 12-3
9.	d	Obj. 12-3
10.	a	Obj. 12-4
11.	b	Obj. 12-4
12.	c	Obj. 12-4

Lesson II		
1.	b	Obj. 12-5
2.	a	Obj. 12-5
3.	c	Obj. 12-5
4.	c	Obj. 12-5
5.	b	Obj. 12-5
6.	b	Obj. 12-5
7.	d	Obj. 12-5
8.	a	Obj. 12-5
9.	b	Obj. 12-6
10.	b	Obj. 12-6
11.	d	Obj. 12-6
12.	a	Obj. 12-6

CHAPTER 13
Ingestive Behavior: Eating

Lesson I: Some Facts About Metabolism and What Starts and Stops a Meal

Read the interim summary on page 381 in your text to re-acquaint yourself with the material in this section.

Learning Objective 13-1 Describe characteristics of the two nutrient reservoirs and the absorptive and fasting phases of metabolism.

Read pages 377-381 and answer the following questions.

1. a. Why do we eat?

   1.   2.

   b. How do our bodies use most of the food we eat?

2. What is the location of the short-term reservoir and what is stored there?

3. Explain how glycogen is produced and the mechanism that stimulates its release.

   a. Cells in the _____ are stimulated by _____, produced in the _____, to convert _____ into glycogen.

   b. What causes the level of glucose in the blood to fall and how is it detected?

   c. What two changes occur in the pancreas in response to a fall in glucose?

   d. What is the effect of glucagon release? (See Figure 13.1 in your text.)

4. a. The short-term carbohydrate reservoir in the liver is the principal fuel supply for what part of the body?

   b. If the short-term reservoir is depleted, what is its next source of reserved fuel?

5. The long-term reservoir of _____ tissue is found beneath the _____ and in the _____ _____. It is filled with _____, complex molecules that contain _____ combined with three _____ _____.

6. What two factors initiate the breakdown of triglycerides into glycerol and fatty acids?

7. Describe the mechanism that saves glucose for the brain during the fasting phase of metabolism. Be sure to mention the role of glucose transporters.

8. To review the fasting phase of metabolism study Figure 13.2 in your text. Continue your review by summarizing the effects of

    a. a fall in the blood glucose level on the pancreas.

    b. the absence of insulin on the cells of the body.

    c. the presence of glucagon on the liver.

    d. and the presence of glucagon and increased activity of the sympathetic nervous system on fat cells.

9. When does the absorptive phase of metabolism begin?

10. What three nutrients are supplied by a well-balanced meal?

11. Describe the changes that occur as each of these nutrients is absorbed.

    a. As carbohydrates break down, what change occurs in the level of glucose in the blood and how is this change detected?

    b. What change occurs in the pancreas as a result in the rise in the level of glucose?

    c. How does insulin affect the cells of the body?

    d. What happens to any extra glucose?

    e. As proteins break down, how are the amino acids used?

    f. How are fats used?

To review the absorptive phase of metabolism again study Figure 13.2 in your text.

12. What observations have suggested to some that total body fat is not regulated?

13. Outline several reasons that suggest that total body fat is somehow regulated.

    a. How may total body fat regulation be affected by

        1. social customs?

        2. an uncertain food supply?

    b. How did food consumption change when

        1. the caloric content of the lunches of human subjects was altered? (Foltin et al., 1990)

        2. animals were force-fed until they became fat? (Hoebel and Teitelbaum, 1966; Steffens, 1975)

Read the interim summary on page 387-388 of your text to re-acquaint yourself with the material in this section.

Learning Objective 13-2 Discuss social and environmental factors that begin a meal and variables that influence dietary selection.

Read pages 381-384 and answer the following questions.

1. Explain why the signals that start and stop a meal are undoubtedly different.

2. a. List factors that encourage us to eat even when we have no physiological need to do so.

    b. Review both classical conditioning and the experiment by describing how Weingarten (1983) studied eating by hungry laboratory rats.

        1. Which stimulus triggered eating behavior?

        2. What do the results suggest about the kind of stimuli that can provoke eating?

    c. What is the relation between meal size and

        1. a fixed meal schedule? (Jiang and Hunt, 1983; de Castro et al., 1986)

        2. a flexible meal schedule? (Le Magnen and Tallon, 1963; 1966; Bernstein, 1981)

        3. the presence of other people? (de Castro and de Castro, 1989)

3. Discuss a significant advantage of being an omnivore as well and the precautions that must be taken.

4. a. Define *sensory-specific satiety* in your own words and then explain how this phenomenon was demonstrated in rats by Le Magnen (1956) and in humans by Rolls et al., (1981).

    b. What is the importance to omnivores of sensory-specific satiety?

5. a. How did two flavored waters given to laboratory rats differ? (Sclafani and Nissenbaum, 1988)

    b. After four days of training, which flavored water did rats prefer? Why? (See Figure 13.4 in your text.)

6. Discuss some of the mechanisms that reduce the likelihood that omnivores will consume foods that are dangerous.

    a. Of what benefit are the receptors on our tongues that detect bitter and sour flavors?

    b. Explain why rats that had tasted saccharin before receiving lithium chloride injections later refused to drink saccharin. Be sure to use the term *conditioned flavor aversion* in your answer. (Garcia and Koelling, 1966)

    c. Briefly explain two situations that confirm that humans too can form conditioned flavor aversions. (Bernstein, 1978)

    d. Discuss how a conditioned flavor aversion can motivate an animal to find the nutrients it needs. (Rozin and Kalat, 1971)

Learning Objective 13-3 Discuss long-term and short-term hunger signals.

Read pages 384-387 and answer the following questions.

1. a. First, a preview of the information to be presented in this section.

    Where are short-term hunger signals detected?

    b. What provides long-term hunger signals?

    c. Under what conditions does the brain become less sensitive to short-term hunger signals? more sensitive?

2. Carefully explain why hunger can be stimulated by an injection of

    a. insulin. What do we call this condition?

    b. 2-deoxyglucose (2-DG). What do we call the condition caused by hypoglycemia or 2-DG?

    c. methyl palmoxirate (MP) or mercaptoacetate (MA). What do we call this condition?

3. Why is less known about the importance of amino acids in the production of hunger signals?

4. a. Describe the experimental procedure that Friedman and his colleagues (Friedman et al., 1986) used to produce moderate glucoprivation and moderate lipoprivation in rats?

    b. Study Figure 13.5 in your text and explain the effects on food intake under each of the experimental conditions.

    c. What do the results suggest about the stimuli that cause hunger?

5. There appear to be two sets of detectors for metabolic fuels. Where are they located and what do they monitor?

6. a. What was the effect on food intake of infusions of 2-DG into the hepatic portal vein? (Novin et al., 1973)

    b. How was this effect abolished?

    c. What do the results suggest about the location of detectors for glucoprivation?

7. a. What is capsaicin?

    b. How did IP injections of capsaicin affect lipoprivic hunger? glucoprivic hunger? (Ritter and Taylor, 1989)

    c. Study Figure 13.6 in your text and explain the subsequent effects on food intake of injections of 2-DG and MA.

    d. What do the results suggest about the location of receptors for glucoprivic hunger?

8. a. How did cutting the vagus nerve just above the abdominal cavity affect lipoprivic hunger? glucoprivic hunger? (Ritter and Taylor, 1990)

   b. What kind of receptors appear to be located in the abdominal cavity and how may they transmit this sensory information to the brain?

9. What stimulus causes lipoprivic detectors in the liver to send hunger signals to the brain?

10. a. How does an injection of 2,5-AM affect glucose metabolism and thus behavior?

    b. How did Tordoff and colleagues (Tordoff et al., 1991) establish that a decrease in metabolic fuels available to the liver causes hunger?

11. Study Figure 13.7 in your text and state how the effects of 2,5-AM were related to the diets of subject rats. (Rawson et al., 1996)

12. _____ is the fuel for cell activity.

13. In addition to 2-DG and 2,5 AM, identify a third drug that causes eating and explain how it does so. (Rawson et al., 1994)

14. Outline the ischymetric hypothesis of hunger proposed by Nicolaïdis. (Nicolaïdis, 1974; 1987)

15. a. How did Ritter et al. (1981) block communication between the third and fourth ventricles of the brain?

    b. When a drug similar in action to 2-DG was then injected into each ventricle how was eating affected?

    c. What is the presumed reason?

16. Finally, summarize evidence that suggests that

    a. hindbrain nutrient receptors may be located in either the area postrema (Bird et al., 1983) or the nucleus of the solitary tract. (Yettefti et al., 1997)

    b. no single set receptors controls eating. (Tordoff et al., 1982; Ritter et al., 1992)

Read the interim summary on page 393 of your text to re-acquaint yourself with the material in this section.

> **Learning Objective 13-4** Discuss the head, gastric, and intestinal factors responsible for stopping a meal.

Read pages 388-393 and answer the following questions.

1. a. Where in the body should we look for the source of short-term and long-term satiety signals?

   b. What effect does this information have on the brain?

2. Rats learn to eat less of a food with a particular flavor that is accompanied by intravenous infusions of glucose. (Mather et al., 1978) What does their behavior suggest about how head factors influence satiety?

3. Define *head factors* in your own words, including the kind of information they detect and their most important role.

4. Why do researchers believe that the stomach is less important in producing hunger and more important in producing satiety? (Ingelfinger, 1944; Davis and Campbell, 1973)

5. a. Explain why Deutsch and Gonzalez (1980) fit rats with pyloric cuffs.

   b. After the rats had fed, the researchers removed 5 ml of food from their stomachs and replaced it with a saline solution. What did they observe? (See Figure 13.8 in your text.)

   c. Carefully explain what the results confirm about the role of the stomach in short-term satiety.

   d. How have other researchers extended this research? (Rauhofer et al., 1983; Seeley et al., 1995)

6. a. Explain why Greenberg and colleagues (Greenberg et al., 1990) attached gastric fistulas to rats. Be sure to refer to sham feeding in your answer.

   b. How was sham feeding affected by an infusion of Intralipid? Intralipid combined with a local anesthetic?

   c. What does the inhibition of sham feeding indicate about the location of short-term satiety signals?

d. Using radioactively labeled Intralipid, what did Greenberg et al. (1991) establish about the timing of the satiating effect?

7. Before studying the role of cholecystokinin (CCK) in satiety, note some information about this hormone.

a. Where and when is it secreted?

b. What is its effect on the gallbladder? the pylorus?

c. How do injections of CCK affect eating? (Gibbs et al., 1973; Smith et al., 1982)

d. Why has it been studied as a satiety signal?

e. What property of CCK restricts the search for its site of action?

8. Summarize evidence that CCK acts peripherally.

a. If the gastric branch of the vagus nerve is cut, how is the suppressive effect of CCK on eating altered? (Smith et al., 1982.)

b. Moran and colleagues (Moran et al., 1989) removed the pyloruses of rats, a region rich in CCK receptors. What change in eating behavior did they observe immediately after surgery and 2-3 months later? (See Figure 13.9 in your text.)

9. Now summarize research to determine how CCK suppresses eating. How did eating or drinking behavior of rats change if

a. an injection of CCK was paired with a particular flavor? (Deutsch and Hardy, 1977)

b. the animals received an injection of an antiemetic drug? (Moore and Deutsch, 1985)

c. after drinking sweetened condensed milk they received low doses of CCK? (Bowers et al., 1992)

d. oxytocin is released in response to an injection of CCK and not by eating a normal meal? (McCann et al., 1989; Stricker and Verbalis, 1991)

10. a. What observation suggested to Russek (1971) that the liver might contain satiety detectors?

b. When he injected glucose into the jugular vein and then the hepatic portal vein, what did he observe?

11. a. How did infusion of small amounts of glucose and fructose into the hepatic portal vein affect rats' appetites for food? (Tordoff and Friedman, 1988)

   b. Why did they conclude that it is the liver that signals satiety to the brain?

12. a. How did injections of either a saline solution or a solution of glucose into the hepatic portal vein of rats who were eating affect their behavior? (Tordoff and Friedman, 1986)

   b. These solutions had been randomly paired with two flavors. Which flavor of food did the rats choose later?

   c. What do these results suggest about the way the brain interprets signals from the liver and the stomach?

13. Finally review research on the mechanisms of long-term body fat regulation.

   a. If an animal gains weight through forced feeding and is later permitted to choose its food, how is subsequent food intake affected? (See Figure 13.10 in your text. Wilson et al., 1990)

   b. If an animal loses weight through enforced dieting, how are satiety factors affected? (Cabanac and Lafrance, 1991)

   c. Study Figure 13.11 in your text and describe the surgical procedure Koopmans (1985) used to channel most of the nutrients eaten by one rat into the body of another.

   d. How did eating behavior and ultimately the affect of satiety signals change as a result?

   e. What are some of the characteristics of the ob mouse? Be sure to mention leptin in your answer. (Campfield et al., 1995; Halaas et al., 1995; Pelleymounter et al., 1995)

   f. If ob mice are given injections of leptin, what physical changes occur? (See Figure 13.12 in your text.)

Lesson I Self Test

1. The short-term fuel reservoir is located in _____ and is filled with _____.

   a. adipose tissue; triglycerides

   b. digestive tract; amino acids
   c. pancreas; glucose
   d. the cells of the liver and muscles; glycogen

2. During the fasting phase of metabolism

a. supplies of glucose are abundant.
b. most cells live on fatty acids.
c. glycerol and fatty acids are converted into triglycerides.
d. excess nutrients are stored in the liver, muscles, and adipose tissue.

3. During the absorptive phase of metabolism

    a. the blood level of glucose rises.
    b. the pancreas ceases to secrete insulin.
    c. proteins, carbohydrates, and fats are used to fuel the cells of the body.
    d. glucose dissolves in fats and is stored in adipose tissue.

4. As recorded in their diaries, subjects ate the most food when

    a. the food was familiar.
    b. the food had a sweet taste.
    c. other people were present.
    d. they were alone.

5. Omnivores

    a. are limited by the distribution of their food.
    b. are completely dependent on one type of food.
    c. do not obtain all essential nutrients from one type of food.
    d. eat only meat.

6. Conditioned flavor aversions

    a. confer an evolutionary advantage on some species and do not occur by chance.
    b. permit omnivores to avoid foods that are dangerous.
    c. can interfere with mechanisms that regulate total body weight.
    d. are short-lived.

7. Lipoprivation can be induced by injections of

    a. insulin.
    b. 2-DG.
    c. capsaicin.

d. methyl palmoxirate (MP).

8. Detectors in the liver that signal lipoprivic hunger appear to be sensitive to

    a. changes in their own internal rate of metabolism.
    b. blood level of particular nutrients.
    c. availability of insulin.
    d. the amount of lipids stored in adipose tissue.

9. Using a pyloric cuff, researchers demonstrated that the

    a. volume of stomach contents is more important than its nutritive content for satiety.
    b. stomach contains receptors that prevent overeating.
    c. stomach contains receptors that monitor the nutritive value of its contents.
    d. stomach communicated information about satiety through the vagus nerve.

10. Injections of cholecystokinin (CCK)

    a. activate stretch receptors in the stomach.
    b. promote stomach emptying.
    c. stimulate a carbohydrate appetite.
    d. suppress eating.

11. By injecting glucose and fructose into the hepatic portal vein, researchers confirmed that the liver

    a. contains receptors that respond when the liver receives nutrients from the intestines.
    b. metabolizes sugars.
    c. is the first organ to signal satiety.
    d. breaks down fatty acids.

12. If ob mice are given an injection of leptin

    a. they lose weight rapidly because they develop diabetes and cannot metabolize glucose.
    b. they eat even greater quantities of food.
    c. their weight returns to normal.
    d. eating behavior is not affected.

Lesson II: Brain Mechanisms of Food Intake and Metabolism and Eating Disorders

Read the interim summary on pages 400-401 of your text to re-acquaint yourself with the material in this section.

Learning Objective 13-5 Describe research on the role of the brain stem and hypothalamus in hunger.

Read pages 393-398 and answer the following questions.

1. Describe research results that demonstrate even decerebrate animals whose brains were transected between the diencephalon and the midbrain, perform ingestive behaviors. How do decerebrate rats respond to

   a. food in their mouths?

   b. different tastes?

   c. hunger and satiety signals?

2. Summarize research on the importance of the AP/NST to hunger.

   a. How is eating and/or FOS production in the AP/NST affected by

      1. injections of 2,5-AM?

      2. severing the branch of the vagus nerve that connects the liver to the brain? (Ritter et al., 1994)

   b. How is lipoprivic hunger and glucoprivic hunger affected by lesions of AP/NST? (See Figure 13.13 in your text. Ritter and Taylor, 1990)

   c. How did injections of glucose or glucagon alter the response of neurons in the NST to a sweet taste? (Giza et al., 1992)

3. a. Where does the AP/NST relay information received from the tongue and internal organs?

   b. How do lesions of the lateral parabrachial nucleus of the pons affect

      1. hunger signals detected by the liver?

      2. lipoprivic feeding and feeding elicited by 2,5-AM? (Calingasan and Ritter, 1993; Grill et al., 1995)

      3. eating produced by 2-DG? What does this finding suggest about neural circuits? (See Figure 13.14 in your text.)

4. Briefly summarize conclusions that prevailed for a long time about the role of the lateral and ventromedial hypothalamus in hunger and satiety. (See Figure 13.15 in your text. Anand and Brobeck, 1951; Teitelbaum and Stellar, 1954; Hetherington and Ranson, 1942)

5. a. Decades later, researchers noted other behavioral changes resulting from these lesions. Describe some of them.

   b. What did Stricker and Zigmond (1976) suggest caused these behavior changes?

6. How did stimulation with injections of excitatory amino acids into the lateral hypothalamus affect eating? injections of a glutamate antagonist? (See Figure 13.16 in your text. Stanley et al., 1993a; Stanley et al., 1996)

7. Let's review more research on the role of the VMH in eating, this time involving the neurotransmitter, neuropeptide Y (NPY).

    a. To begin, how does NPY affect food intake? (Clark et al., 1984) How do infusions of this substance into the VMH affect eating? Be sure to mention the kinds of behaviors that rats will engage in to obtain food. (Flood and Morley, 1991; Jewett et al., 1992)

    b. What changes result from infusions of NPY into the midlateral hypothalamus? (Stanley et al., 1993b) into the paraventricular nucleus in the medial hypothalamus? (Wahlestedt et al., 1987; Abe et al., 1989; Currie and Coscina, 1996)

    c. Hypothalamic levels of NPY are increased by _____ _____ and decreased by _____. (Sahu et al., 1988)

    d. If NPY receptors are blocked, how is eating affected? (Myers et al., 1995)

    e. What does this last finding suggest about NPY and normal eating?

8. a. Where are neurons that secrete NPY located? (Look back at Figure 13.15.)

    b. Where do these NPY-containing neurons send dense projections? lighter projections? Cite research that the lighter projections play a role in hunger and the control of metabolism. (Bai et al., 1985; Akabayashi et al., 1994)

9. a. What is the effect of 2-DG induced glucoprivation on the

    1. NPY-secreting neurons in the arcuate nucleus?

    2. production of Fos protein in both the arcuate nucleus and the paraventricular nucleus? (See Figure 13.17 in your text. Minami et al., 1995)

    b. Cite research that these neurons respond only to glucoprivation detected on the brain side of the blood-brain barrier. (Akabayashi et al., 1993)

10. a. Briefly summarize some of the other effects of NPY. How does NPY affect

    1. energy expenditure? (Egawa et al., 1991)

2. ovulation and sexual behavior? (Clark et al., 1985)

b. Explain how temporary infertility conserves energy. (Wade et al., 1996)

11. a. What kind of nutrient do rats prefer during their first big meal? (Leibowitz et al., 1988)

b. Which neurotransmitter and which brain structure may play an important role in carbohydrate appetite?

12 a. What relationship was observed between carbohydrate intake and

1. microinfusions of norepinephrine (NE) into the PVN? (See Figure 13.18 in your text. Leibowitz et al., 1985)

2. infusions into the PVN of an NE agonist and antagonist? (Yee et al., 1987)

3. lesions of the PVN or destruction of the noradrenergic axons that enter it? (Shor-Posner et al., 1986)

4. severing the vagus nerve that serves the pancreas and introducing NE into the PVN? (Sawchenko et al., 1981)

5. recordings of NE levels across the sleep-waking cycles? (See Figure 13.19 in your text. Stanley et al., 1989)

b. Galanin, a _____, is found with _____ in the terminal buttons in the _____ and the two are released _____. (Melander et al., 1987)

c. How do infusions of galanin into the PVN affect eating and nutrient preference? (Tempel et al., 1988)

d. And how do infusions of galanin affect the secretion of insulin and corticosterone? (Tempel and Leibowitz, 1990)

e. Compare the galanin levels in the PVN of rats with a natural preference for low-fat and high-fat diets. (Study Figure 13.20 in your text. Akabayashi et al., 1994)

f. What kind of neurons do galanin-secreting neurons form synapses with? (Horvath et al., 1996)

Learning Objective 13-6 Describe research on the role of the hypothalamus on satiety

Read pages 398-400 and answer the following questions.

1. Restate the most striking behavior of animals with a lesion of the ventromedial hypothalamus (VMH) and the long accepted explanation for the VMH syndrome.

2. Other aspects of eating are also affected. How is food intake affected if these animals

    a. are fed a diet to which quinine has been added? (Ferguson and Keesey, 1975)

    b. given a choice of different diets? (Sclafani and Aravich, 1983)

3. Carefully explain how VMH lesions disrupt the control of the autonomic nervous system and why food intake is affected. (Weingarten et al., 1985)

4. VMH lesions destroy not only the ventromedial hypothalamus but also axons that connect the paraventricular nucleus of the hypothalamus (PVN) with structures in the brains stem. According to Kirchgessner and Sclafani (1988), why might the destruction of these axons cause overeating?

5. a. How do injections of 5-HT (serotonin) into the PVN, VMH, and SCN affect carbohydrate appetite? (Leibowitz et al., 1990)

    b. For this effect to occur, during what portion of the day-night cycle must injections be given?

    c. What explanation of the results do the researchers suggest?

6. a. What is the general effect of drugs that destroy serotonergic neurons, inhibit the synthesis of 5-HT, or block 5-HT receptors? (Breisch et al., 1976; Saller and Stricker, 1976; Stallone and Nicolaïdis, 1989)

    b. How do injections of a 5-HT antagonist affect NPY production and food intake? (Dryden et al., 1995)

7. a. How has research on 5-HT and carbohydrate intake been used to benefit obese people? Be sure to mention fenfluramine (FEN) and it affect on appetite in your answer.

    b. What region is not a site of action for FEN? (Fletcher et al., 1993) What region may be? (Li et al., 1994)

8. a. Once again, where is leptin secreted? How did infusions of leptin into the cerebral ventricles affect eating and production of NPY in the arcuate nucleus? (Schwartz et al. 1996.)

    b. Where do the NPY-secreting neurons in the arcuate nucleus project?

    c. What kind of receptors are found in the NPY-secreting neurons in the PVN? (Hakansson et al., 1996; Mercer et al., 1996. See Figure 13.21 in your text.)

    d. How do infusions of glutamate affect the NPY-secreting neurons of the arcuate nucleus? infusions of leptin? (Study Figure 13.22 in your text. Glaum et al., 1996)

e. In what range is the body weight of mice who cannot produce either leptin or NPY because of natural and targeted mutations? What do these results suggest about the roles of both of these substances? (Erickson et al., 1996)

Read the interim summary on page 408 of your text to re-acquaint yourself with the material in this section.

Learning Objective 13-7 Discuss the physiological factors that may contribute to obesity.

Read pages 401-404 and answer the following questions.

1. a. What may be an unfortunate consequence of urging children to "Clean up your plate?" (Birch et al., 1987)

   b. What other social customs also contribute to this problem?

2. a. What was the long-term success rate for people who had participated in a behavioral weight loss program? (Kramer et al., 1989)

   b.. According to Wooley and Garner (1994), what should we recognize about weight loss programs and what should be done?

3. Refute these misconceptions about obesity.

   a. "Rhonda is so fat. I bet she snacks on fast food all the time."

   b. Unhappiness and depression cause obesity. (Rodin et al., 1989)

   c. Obese people would loose weight if they would only stick to their diets.

4. a. When Lichtman et al. (1992) compared their direct measurements of the food intake and physical activity of subjects in a controlled environment with the subjects' self-reports, what did they learn?

   b. How did the subjects react to these findings?

5. List the two ways that we expend energy. Underline the form that accounts for most of our energy expenditure. (Calles-Escandon and Horton, 1992)

   1.   2.

6. Explain how metabolic efficiency differs in obese and nonobese people. (Rose and Williams, 1961)

7. How did Sims and Horton (1968) study the response of nonobese men to overeating?

8. a. According to twin studies, approximately how much of the variability in body fat is due to genetic differences? (Price and Gottesman, 1991; De Castro, 1993; Allison et al., 1996)

    b. List some of the determinants of body weight that may be affected by heredity. (Bouchard, 1989, 1991)

    1.   3.

    2.   4.

    c. How did the body weight of a sample of people who had been adopted correlate with

    1. their biological and adoptive parents? (Stunkard et al., 1986)

    2. their full or half siblings with whom they had not been raised? (Sørensen et al., 1989)

    d. What do the results of these studies suggest about a hereditary basis of metabolism?

9. a. State James and Trayhurn's (1981) hypothesis concerning the origin of genetic differences in metabolic efficiency.

    b. Summarize the conclusions of research with the inhabitants of Nauru (Gibbs, 1996) and two groups of Pima Indians (Ravussin et al., 1994) that support James and Trayhurn's hypothesis.

10. Let's look again at the role of leptin in human obesity. Is there a relationship between plasma levels of leptin and

    a. the total body fat of obese and lean people? (Schwartz et al., 1996)

    b. the body weight of Pima Indians? (Ravussin et al., 1997)

11. a. If leptin plays a role in human obesity, the mechanism may be reduced _____ to the hormone and not decreased _____.

    b. Explain how leptin insensitivity may result from

    1. a faulty transport system across the blood-brain barrier. (Caro et al., 1996)

    2. a mutation of the gene responsible for the production of the leptin receptor. (Gura, 1997)

12. a. How does a genetic defect affect the agouti mouse? Be sure to mention the agouti protein and its role in your answer.

    b. What is the effect on eating?

1. if melanocortin-4 receptors (MC4-R) are stimulated with melanocortin?

2. if MC4-receptor agonists are injected into the cerebral ventricles? MC4-receptor blockers? (Fan et al., 1997)

3. of a targeted mutation of the gene responsible for the production of the MC4 receptor? (Huszar et al., 1997)

c. Why may an understanding of the function of the MC4 receptor benefit humans?

13. Finally, list some additional genetic defects that may be involved in obesity. (Roberts and Greenberg, 1996; Clément et al., 1995)

1.

2.

3.

4.

5.

Learning Objective 13-8 Discuss the mechanical, surgical, and pharmacological treatments of obesity.

Read pages 404-405 and answer the following questions.

1. a. When people who have had their jaws wired together to help them loose weight have the wiring removed, what almost always happens?

b. In an attempt to help patients maintain their losses, what did some therapists do?

c. And what did about half their patients eventually do?

2. List the two kinds of surgical procedures for reducing food intake.

1.   2.

3. a. Describe gastroplasty, the most common procedure.

b. Following surgery, how should patients feel after eating a small amount of food? Unfortunately, how do they often feel? (Be sure to mention *nimiety* in your answer.)

4. Briefly summarize some of the common side effects of intestinal bypass surgery.

5. a. What is the "Gastric Bubble" and why did some therapists suggest it for their patients?

b. Why was it later taken off the market in the United States? (Kral, 1989)

6. Study Figure 13.23 in your text and describe the weight changes of 176 patients while they were taking fenfluramine and after they stopped taking it. (Be sure that you understand the effect of fenfluramine in the brain. Bray, 1992)

7. According to Bray, what may be the reason regulatory agencies and boards are hesitant to approve of long-term drug treatment for obesity?

8. Table 13.1 in your text lists some of the anti-obesity drugs currently being developed by drug companies.

> **Learning Objective 13-9** Discuss the physiological factors that may contribute to anorexia nervosa and bulimia nervosa.

Read pages 405-408 and answer the following questions.

1. Describe the symptoms of anorexia nervosa.

2. Now describe the symptoms of bulimia nervosa and its aftermath. (Mawson, 1974; Halmi, 1978)

3. a. Study Figure 13.24 in your text and compare the effects of the sight and smell of warm cinnamon rolls on the insulin levels of anorexic and lean, but not anorexic, young women. (Broberg and Bernstein, 1989)

    b. What do the results indicate about the interest of anorexics in food?

4. a. To combat their intense fear of becoming obese, what do many anorexics do?

    b. What do studies with animals suggest may be a reason for increased exercise? (Routtenberg, 1968; Wilckens et al., 1992)

5. Which explanation for anorexia nervosa—biological or social—is favored by most psychologists?

6. a. Twenty years after treatment for anorexia, what percentage of a group of patients showed good recovery? (Ratnasuriya et al., 1991)

    b. What had happened to almost 15 percent of them?

7. How does anorexia affect

    a. bone density?

    b. menstruation?

    c. the brain? (Artmann et al., 1985; Herholz, 1996; Kingston et al., 1996)

8. a. What do twin studies suggest about the cause of anorexia? (Russell and Treasure, 1989; Walters and Kendler, 1995)

    b. Summarize the biochemical changes in the brain associated with anorexia and bulimia reported in a literature review by Fava et al. (1989).

    c. Why are endocrine system changes probably effects rather than causes of this disorder?

9. a. Describe the levels of neuropeptide Y in the cerebrospinal fluid of patients with severe anorexia and the same patients after they have regained their normal weight. (Kaye et al., 1990; Kaye, 1996)

    b. What symptom of anorexia may be affected by this neuropeptide?

10. a. Which kind of drugs have been found to be ineffective in treating anorexia? (Mitchell, 1989)

    b. Summarize the limited success of treatment with cyproheptadine. (Halmi et al., 1986)

    c. What class of drugs may be useful in treating bulimia nervosa? (Kennedy and Goldbloom, 1991; Advokat and Kutlesic, 1995)

Lesson II Self Test

1. Lesions of the area postrema and nucleus of the solitary tract (AP/NST)

    a. stimulate a carbohydrate appetite.
    b. reduce the ability to distinguish between flavors.
    c. stimulate fos production.
    d. abolish both glucoprivic and lipoprivic feeding.

2. Lesions of the _____ produce _____ and lesions of the _____ abolish _____.

    a. lateral hypothalamus; hunger; ventromedial hypothalamus; satiety
    b. ventromedial hypothalamus; overeating; lateral hypothalamus; eating.
    c. paraventricular nucleus; overeating; ventromedial hypothalamus; undereating
    d. ventromedial hypothalamus; satiety; paraventricular nucleus; obesity

3. Neuropeptide Y , which is secreted by neurons in the _____, _____.

    a. area postrema; controls hormones that regulate the fasting phase of metabolism.
    b. paraventricular nucleus; causes a rapid decline in blood glucose levels.
    c. arcuate nucleus; stimulates ravenous eating.
    d. ventromedial hypothalamus; abolishes eating.

4. Norepinephrine _____ carbohydrate intake and serotonin _____ it.

    a. abolishes; stimulates
    b. increases; decreases
    c. stimulates; increases
    d. decreases; increases

5. Leptin is _____

a. secreted by well-fed adipose tissue and inhibits eating.

b. a by-product of protein metabolism and inhibits the release of neuropeptide Y.

b. secreted by the liver during the fasting phase of metabolism and decreases metabolic rate.

d. secreted by neurons in the brain and does not cross the blood-brain barrier.

6. Which one of these conditions appears to contribute least to overeating and obesity?

a. social customs concerning food
b. unhappiness and depression
c. hereditary differences
d. an efficient metabolism

7. People with an efficient metabolism

a. must eat more food to maintain their body weight.

b. have difficulty loosing weight even on a reduced calorie diet.

c. have difficulty matching food intake to physical activity.

d. do not have any calories left over for deposit in long-term nutrient reservoirs.

8. The agouti mouse has a mutation of the gene responsible for the production of _____ receptors.

a. neuropeptide Y (NPY)
b. leptin

c. melanocortin-4
d. galanin

9. Gastroplasty, a surgical procedure to help obese people loose weight, results in

a. feelings of satiety when a small amount of food is eaten.

b. nausea when too much food is eaten.

c. increased secretion of CCK

d. nimiety when a small amount of food is eaten.

10. Fenfluramine, a drug that helps obese people loose weight, stimulates the release of

a. neuropeptide Y.
b. insulin.
c. serotonin.
d. norepinephrine.

11. Anorexics

a. are unresponsive to the effects of food.
b. do not experience hunger.
c. have an intense fear of becoming obese.
d. attempt to reduce their need for calories by reducing physical activity.

12. The cerebrospinal fluid of anorexics contains elevated levels of

a. neuropeptide Y.
b. cholecystokinin (CCK).
c. fenfluramine.
d. galanin.

Answers for Self Tests

Lesson I

1. d Obj. 13-1
2. b Obj. 13-1
3. a Obj. 13-1
4. c Obj. 13-2
5. c Obj. 13-2
6. b Obj. 13-2
7. d Obj. 13-3
8. a Obj. 13-3
9. c Obj. 13-3
10. d Obj. 13-4
11. a Obj. 13-4
12. c Obj. 13-4

Lesson II

1. d Obj. 13-5
2. b Obj. 13-5
3. c Obj. 13-5
4. b Obj. 13-6
5. a Obj. 13-6
6. b Obj. 13-7
7. b Obj. 13-7
8. c Obj. 13-7
9. d Obj. 13-8
10. c Obj. 13-8
11. c Obj. 13-9
12. a Obj. 13-9

CHAPTER 14
Learning and Memory: Basic Mechanisms

Lesson I: The Nature of Learning, Learning and Synaptic Plasticity, and Perceptual Learning

Read the interim summary on page 414 of your text to re-acquaint yourself with the material in this section.

Learning Objective 14-1 Describe four of the basic forms of learning—perceptual learning, stimulus-response learning, motor learning, and relational learning.

Read pages 410-414 and answer the following questions.

1. _____ physically change the structure of our _____ _____ and thereby change our _____. This process is called _____ and these changes are called _____.

2. State the primary function of the ability to learn.

3. List four of the basic forms of learning.

   1.   3.

   2.   4.

4. Define *perceptual learning* in your own words.

5. State its primary function.

6. a. How many of our sensory systems are capable of perceptual learning?

   b. Where does perceptual learning appear to take place?

7. Define *stimulus-response learning* in your own words, noting its two major categories.

8. Which neural circuits are presumably involved?

9. In your own words, briefly explain what happens during classical conditioning.

10. Let's examine how the species-typical defensive eyeblink response of a rabbit can be conditioned to a tone. Identify the

    a. unconditional stimulus (US)   c. conditional stimulus (CS)

    b. unconditional response (UR)   d. conditional response (CR)

11. Study Figure 14.1 in your text and then complete Figure 1, which illustrates the kinds of changes that may take place in the brain during classical conditioning.

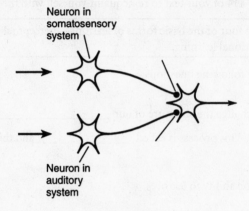

Figure 1

12. State the Hebb rule in your own words.

13. Briefly explain what happens during instrumental conditioning. Be sure to use the terms *reinforcing stimuli* and *punishing stimuli* in your answer.

14. Study Figure 14.2 in your text and then complete Figure 2 on the next page, which illustrates the kinds of changes that may take place in the brain during instrumental conditioning.

15. Now that you have described classical and instrumental conditioning, discuss three ways in which they differ.

16. Define *motor learning* in your own words, and explain why it is considered a form of stimulus-response learning. (Study Figure 14.3 in your text.)

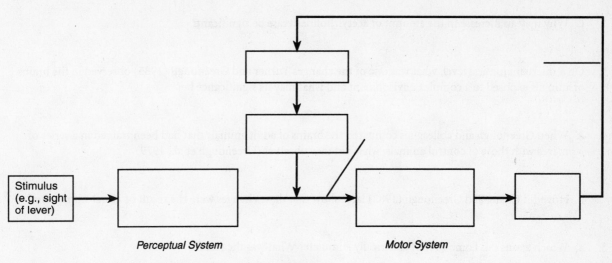

Figure 2

17. Define *relational learning* in your own words.

18. Briefly describe these forms of relational learning.

   a. spatial learning

   b. episodic learning

   c. observational learning

Read the interim summary on pages 426-427 of your text to re-acquaint yourself with the material in this section.

> *Learning Objective 14-2* Discuss research on how learning affects neural structures, the induction of long-term potentiation, and the role of NMDA receptors.

Read pages 414-421 and answer the following questions.

1. Define *synaptic plasticity* in your own words.

2. a. Why did Rosenzweig and his colleagues divide litters of laboratory rats and place them in impoverished and enriched environments? Be sure to describe both kinds of environments. (Weiler et al., 1995; Rosenzweig and Bennett, 1996)

   b. What were some of the many differences they observed in the brains of animals in the enriched environment?

c. Why may an increase in the amount of acetylcholinesterase be significant?

3. On a microanatomical level, what was one of the changes Turner and Greenough (1985) observed in the brains of animals exposed to a complex environment and what may its significance be?

4. a. When Greenough and colleagues compared the brains of adult animals that had been trained in a series of mazes with those of control animals, what did they observe? (Greenough et al., 1979)

    b. How did Chang and Greenough (1982) later show that these changes were the result of visual stimulation?

5. a. Which axons did Lømo (1966) electrically stimulate? What was the result?

    b. What is this increase called?

6. a. Return to Figure 3.13 in your text and review the location of the hippocampus in the human brain.

    b. List the structures that comprise the hippocampal formation.

    1.   3.

    2.   4.

    c. Underline the name of the structure through which the primary inputs and outputs of the hippocampal formation pass.

7. Let's trace the pathway of incoming information through the hippocampal formation.

    a. Where do neurons in the entorhinal cortex relay incoming information?

    b. Where, in turn, does this structure send axons?

    c. What are the two major divisions of the hippocampus proper? Which one of these receives axons from the dentate gyrus?

    d. Briefly describe how an axon grows from a pyramidal cell in field CA3.

    e. Why are dendritic spines important for long-term potentiation?

    f. CA3 axons branch in two directions. Where does each of these branches travel?

    g. Finally, where do the pyramidal cells of field CA1 send their axons?

To review this pathway study Figure 14.4 in your text.

8. Study Figures 14.5 and 14.6 in your text and explain a typical procedure for producing long-term potentiation.

    a. Where is a stimulating electrode placed? a recording electrode?

b. Define *population EPSP* in your own words.

c. What event triggers a population EPSP and what does the size of the first one indicate?

d. How is long-term potentiation induced?

e. What evidence confirms that long-term potentiation has occurred?

9. a. Briefly describe how long-term potentiation is induced in isolated slices of the hippocampal formation. (See Figure 14.7 in your text.)

    b. What are some of the advantages of this procedure?

10. Define *associative long-term potentiation* in your own words.

11. Study Figures 14.8 and 14.9 in your text and explain the procedure Kelso and Brown (1986) use to produce it.

    a. Which inputs were stimulated?

    b. Which synapses were strengthened?

    c. What do the results suggest about the selective effects of stimulation?

    d. How does this phenomenon appear to confirm the Hebb rule?

12. Explain why long-term potentiation occurs if

    a. a series of pulses is given at a high rate all in one burst, but not if the same number of pulses are delivered at a slow rate.

    b. artificially depolarized axons are stimulated, but not if depolarization and stimulation occurred at different times. (See Figure 14.10 in your text. Kelso et al., 1986)

13. To review: What two events are necessary if long-term potentiation is to occur?

    1.   2.

14. Study Figure 14.11 in your text and explain the role of NMDA receptors in long-term potentiation.

    a. Where are NMDA receptors found?

b. What kind of ion channel does an NMDA receptor control?

c. Carefully explain what two conditions must occur in order for calcium to enter cells through the ion channels controlled by NMDA receptors?

15. a. Briefly explain why research with drugs such as AP5 that block NMDA receptors supports their importance in long-term potentiation. (Brown et al., 1989)

    b. Which receptors are responsible for synaptic transmission in the newly strengthened synapses?

16. a. How did an injection of EGTA affect

    1. hippocampal pyramidal cells at the injection site?

    2. long-term potentiation in these cells and neighboring cells? (Lynch et al., 1984)

    b. What do these results suggest about the role of calcium in long-term potentiation?

17. Study Figure 14.12 in your text and explain how researchers increased the amount of calcium in cells and how long-term potentiation was affected. (Malenka et al., 1988)

18. Let's look now at the role of dendritic spikes that occur in some kinds of pyramidal cells during long-term potentiation. Carefully explain

    a. what triggers dendritic spikes.

    b. how the membrane potential of the cell and dendrites then changes.

    c. and how the amount of calcium in the dendrites changes. (See Figure 14.13 in your text. Yuste and Denk, 1995.)

19. Using TTX, which prevents the formation of dendritic spikes, what did researchers confirm about the necessary conditions for long-term potentiation? (See Figure 14.14 in your text. Magee and Johnston, 1977)

20. Finally, study Figure 14.15 in your text and carefully explain why both weak and strong synapses on a dendritic spine must be active at the same time in order for ion channels controlled by NMDA receptors to open and associative long-term potentiation to occur.

Learning Objective 14-3 Discuss the mechanisms responsible for the increase in synaptic strength that occurs during long-term potentiation.

Read pages 421-426 and answer the following questions.

1. Outline the changes that may be responsible for synaptic strengthening. (See Figure 14.16 in your text.)

2. Let's look at these possibilities more carefully.

    a. How did Tocco et al., (1992) establish long-term potentiation in intact laboratory animals?

    b. Describe how brain slices were prepared for histological examination and the results of the examination.

    c. What do the results suggest about the mechanism of synaptic strengthening?

3. When researchers controlled whether NMDA receptors were blocked by the magnesium ion, what kind of receptors increased in field CA1, presumably as a result of long-term potentiation? (Liao et al., 1995)

4. a. Long-term potentiation begins _____ with the entry of _____ ions into the dendrites, which activates special _____-_____ _____, called _____ _____.

    b. Once activated, how do protein kinases affect protein molecules?

5. How does the presence of CaM-KII or tyrosine kinase in post-synaptic neurons appear to affect the successful production of long-term potentiation? (Silva et al., 1992a; Grant et al., 1992)

6. a. Identify two changes that occurred in a subpopulation of small spines on the postsynaptic dendrite following long-term potentiation by studying Figure 14.17 in your text. (Hosokawa et al., 1995)

    b. What term is used to refer to these altered synapses?

    c. How did Buchs and Muller (1996) demonstrate that perforated synapses are a characteristic feature of strengthened synapses?

7. Briefly summarize Edwards' (1995) explanation of why the number of postsynaptic AMPA receptors increases. (Study Figure 14.18 in your text.)

8. a. How is long-term potentiation affected if anisomycin, a drug that block protein synthesis, is administered

    1. before, during, or immediately after a burst of stimulation?

2. one hour after stimulation? (Frey et al., 1988)

b. What, then, is a necessary condition if longlasting long-term potentiation is to occur?

9. Discuss evidence that suggests that protein synthesis takes place in the dendrites. (Tiedge and Brosius, 1996)

10. a. By what means might enzyme-initiated changes in the dendritic spines affect the nearby terminal buttons? Be sure to use the term *retrograde messenger* in your answer.

b. Why are nearby terminal buttons the only ones affected?

c. How do drugs that block the production of nitric oxide affect the establishment of long-term potentiation? (For example, O'Dell et al., 1991; Schuman and Madison, 1991)

To review the chemical reactions that occur during long-term potentiation study Figure 14.19 in your text.

11. a. Define *long-term depression* in your own words.

b. How does AP5 affect the establishment of long-term potentiation and long-term depression? (See Figure 14.20 in your text. Dudek and Bear, 1992)

c. Describe several patterns of electrical stimulation that produce long-term depression. (Stanton and Sejnowski, 1989; Debanne et al., 1994; Thiels et al., 1996)

d. What do these findings suggest about the Hebb rule and learning?

12. a. List some areas of the brain other than the hippocampal formation where long-term potentiation has been successfully demonstrated.

b. What are some of the implications of finding that long-term potentiation can be produced in several different brain locations?

Read the interim summary on pages 434-435 of your text to re-acquaint yourself with the material in this section.

Learning Objective 14-4 Describe research on the role of the primary visual cortex in visual perceptual learning.

Read pages 427-432 and answer the following questions.

1. Define *perceptual learning* in your own words.

2. Review the divisions and flow of information in visual association cortex by studying Figure 14.21 in your text.

3. a. List the two divisions of the second level of visual association cortex.

1.   2.

b. What is their common beginning? Where does each continue?

c. What is the function of each in visual perception?

4. What kind of brain damage disrupted the monkeys' ability to perform a visual discrimination task? Why? (Study Figure 14.22 in your text. Mishkin, 1996)

5. What kind of stimuli may neurons near the superior temporal sulcus detect? (Baylis et al., 1985; Rolls and Baylis, 1986)

6. How did Rolls et al., (1989) demonstrate changes in the response patterns of neurons in inferior temporal cortex when new stimuli are presented?

7. Use one word to describe the capacity of visual perceptual learning.

8. a. Describe the visual displays that Moscovitch and his colleagues asked people to memorize. (See Figure 14.23 in your text. Moscovitch et al., 1995)

b. Now describe the two retrieval tasks they used to test recall.

c. Study Figure 14.24 in your text and summarize testing results.

d. What do the findings suggest about the location for recall of perceptual memories for identities and locations of objects?

9. Define *short-term memory* in your own words.

10. a. Study Figure 14.25 in your text and describe the delayed matching-to-sample procedure that Fuster and Jervey (1981) used to test short-term memory.

b. Study Figure 14.26 in your text which shows the response of a single representative neuron during this task.

  1. Describe the response of this neuron to red light and green light.

  2. What are the implications of the sustained response to red light?

11. a. If neurons in the inferior temporal cortex are stimulated during the delay interval of a delayed matching-to-sample task, how is recall affected? (Kovner and Stamm, 1972)

b. What do the results suggest about the conditions necessary for visual short-term memory?

12. a. How did Oyachi and Ohtsuka (1995) temporarily disrupt the activity of human posterior parietal cortex?

    b. How did this disruption affect the performance on a delayed matching-to-sample task?

    c. What do the results suggest about the function of this brain region?

13. a. Let's look at other brain regions that are involved in short-term memory. Where do the major regions of visual association cortex form connections?(Wilson et al., 1993)

    b. If the activity of the dorsolateral prefrontal cortex is disrupted, recognition of what kinds of stimuli are affected?

    c. According to most researchers, why may the dorsolateral prefrontal cortex play a role in the recognition of sensory stimuli?

14. a. Describe the delayed responding task Quintana and Fuster (1992) taught monkey subjects.

    b. During testing, what kinds of information did some neurons in dorsolateral prefrontal cortex appear to encode? other neurons?

    c. What do the results suggest about the aspect of a task subjects remember?

Learning Objective 14-5 Describe research on the role of acetylcholine in auditory learning.

Read pages 432-434 and answer the following questions.

1. How do we use the wealth of information we obtain and store through perceptual learning?

2. What kind of events appear to facilitate perceptual learning? which neurotransmitter?

3. What may be the cause of the memory loss that is one of the first symptoms of Alzheimer's disease? Cite research to support your answer. (Nakano and Hirano, 1982; Whitehouse et al., 1982)

4. a. Outline how researchers classically conditioned an emotional response in guinea pigs. (Bakin and Weinberger, 1990; Edeline and Weinberger, 1991a; 1991b; 1992)

b. Study Figure 14.27 in your text and compare the response pattern of neurons in the auditory system before and after training.

c. What do the results suggest about the effects of auditory learning on neurons in the auditory system?

d. How long do these changes last?

e. Study Figure 14.28 in your text and explain the important role of acetylcholine in this kind of leaning. Be sure to refer to the nucleus basalis in your answer.

5. a. If the nucleus basalis is stimulated at the same time a tone of a particular frequency is presented, how is the response of the neurons in the auditory cortex affected? Why? (Bakin and Weinberger, 1996)

b. How can this effect be blocked?

c. How did Cruikshank and Weinberger (1996) demonstrate a similarity between this response and long-term potentiation?

d. What do their results indicate about an important element of classical conditioning?

Lesson I Self Test

1. In classical conditioning, the unconditional stimulus

   a. always elicits the species-typical response.
   b. is a neutral stimulus.
   c. elicits the species-typical response if it has previously been paired with the conditional stimulus.
   d. initially has little effect on behavior.

2. Instrumental conditioning results from an association between

   a. two stimuli.
   b. a stimulus and a response.
   c. a conditional and an unconditional stimulus.
   d. two responses.

3. The Hebb rule states that a synapse will be strengthened if it repeatedly become active _____ the _____ neuron fires.

   a. at the same time; presynaptic
   b. soon after; presynaptic
   c. about the same time; postsynaptic
   d. before; postsynaptic

4. In order for long-term potentiation to occur

   a. the presynaptic membrane must be depolarized at the same time that the synapses are active.
   b. the postsynaptic membrane must be depolarized at the same time that the synapses are active.
   c. weak and strong synapses to a single neuron must be stimulated at approximately the same time.
   d. a series of electrical pulses must be delivered at a slow rate.

5. The size of the

   a. population EPSP was predicted by the Hebb rule.

b. first population EPSP indicates the strength of synaptic connections before long-term potentiation takes place.

c. population EPSP decreases if long-term potentiation has taken place.

d. population EPSP slowly increases for up to 40 hours if long-term potentiation has taken place.

6. NMDA receptors

a. control calcium ion channels.

b. are found in highest concentration in the mossy fibers of field CA3.

c. detect the presence of magnesium.

d. are blocked by glutamate.

7. One of the effects of long-term potentiation is a(n)

a. increase in the number of calcium-dependent enzymes.

b. increase in the amount of postsynaptic thickening of neurons in the hippocampal formation.

c. decrease in protein synthesis in the cell body.

d. increase in the number of postsynaptic AMPA receptors.

8. Nitric oxide

a blocks the effects of calcium.

b. may increase the number of glutamate receptors.

c. may be produced in the dendritic spines.

d. lasts for a long time and can diffuse the entire length of a postsynaptic axon.

9. Long-term depression may result from

a. sustained increases in protein synthesis in the cell body.

b. the gradual atrophy of the dendrites that transmit chemical messages to the cell.

c. low-frequency stimulation of the synaptic inputs to a cell.

d. increased sensitivity of protein kinases to calcium.

10. In a delayed matching-to-sample task, the "delay" is the interval

a. between teaching the subject the task and testing to see whether it has learned it.

b. between the sample stimulus and the choices.

c. between successive trials.

d. a subject must pause before responding.

11. In a delayed matching-to-sample task, neurons in inferior temporal cortex continued to respond during the delay interval, which suggests that these neurons are parts of circuits

a. involved in instrumental conditioning.

b. responsible for reinforcement.

c. involved in excitation rather than inhibition.

d. that remember that a particular stimulus was presented.

12. The release of _____ from neurons in the _____ causes neurons in the auditory cortex _____.

a. acetylcholine; nucleus basalis; to become more sensitive to the auditory input they are receiving

b. glutamate; medial geniculate nucleus of the thalamus; to fail to differentiate between tones of different frequencies

c. calcium; NMDA receptors; to respond more vigorously to the presence of calcium

d. nitric oxide synthase; presynaptic neuron; to increase protein synthesis necessary for long-term potentiation.

Lesson II: Classical Conditioning, Instrumental Conditioning and Motor Learning

Read the interim summary on pages 436-437 in your text to re-acquaint yourself with the material in this section.

Learning Objective 14-6 Discuss the physiology of the classically conditioned emotional response to aversive stimuli.

Read pages 435-436 and answer the following questions.

1. A hypothetical neural circuit of the changes in synaptic strength resulting from a classically conditioned emotional response is diagrammed in Figure 14.29 in your text.

a. In this example, what is the CS? the US?

b. Where is information about the CS sent? the US?

c. Where does information from the CS and US converge?

d. When a rat does encounter a painful stimulus that is paired with a tone, how are the synapses in the MGm and basolateral amygdala strengthened according to the Hebb rule?

Let's look at experimental support that the basolateral amygdala and the MGm are involved in learning.

2. What kind of lesions disrupt conditioned emotional responses established with a tone followed by a foot shock? (Iwata et al., 1986; LeDoux et al., 1986, 1990; Sananes and Davis, 1992)

3. What evidence suggests synaptic changes do occur in the MGm? (reported by Weinberger, 1982)

4. a. Summarize the changes in neurons in the lateral amygdala recorded during pairings of a tone with a footshock. (Study Figure 14.30 in your text. Quirk et al., 1995)

b. Explain why these researchers believe the amygdala to be a more important site for learning than the MGm?

5. a. What is the result of stimulation of the axons that bring auditory information to the MGm (Gerren and Weinberger, 1983) and those that connect the medial geniculate nucleus to the lateral amygdala (Clugnet and LeDoux, 1990)?

b. Following long-term potentiation of the lateral nucleus of the amygdala, what changes are seen in the response of neurons in this nucleus to auditory stimuli? (Rogan and LeDoux, 1995)

6. When researchers injected AP5 into the basolateral amygdala after establishing a classically conditioned emotional response, was the response affected? What do the results suggest about the role of NMDA receptors? (Campeau et al., 1992; Fanselow and Kim, 1994)

7. a. Define *extinction* in your own words.

b. If, however, AP5 is injected into the amygdala just before extinction training takes place, is extinction affected? (Falls et al., 1992)

Read the interim summary on page 449 to re-acquaint yourself with the material in this section.

> **Learning Objective 14-7** Describe the role of the basal ganglia and premotor cortex in instrumental conditioning and motor learning.

Read pages 437-441 and answer the following questions.

1. To review: Instrumental conditioning involves the strengthening of connections between _____ _____ that detect a particular _____ and those that produce a particular _____. These _____ _____ begin in various regions of _____ _____ cortex and end in _____ _____ cortex. The two major pathways between them are direct _____ connections and connections through the _____ _____ and the _____.

2. Describe some of the functions of each of these pathways along with a task that illustrates each function.

3. Discuss how we learn a complex behavior through instrumental conditioning and the effects of practice on the basal ganglia and transcortical circuits.

4. a. Which nuclei comprise the neostriatum?

    b. Which regions send information to them and where do they, in turn, send information? (See Figure 14.31 in your text.)

    c. What kind of lesions do and do not disrupt a visual discrimination task? (Divac et al., 1967; Gaffan and Harrison, 1987; Gaffan and Eacott, 1995)

5. a. List the kinds of learning tasks that McDonald and White (1993) taught laboratory rats.

       1.

       2.

       3.

    b. Next to each kind of learning, note the location(s) of lesions that disrupted it.

    c. What do the results suggest about the role of particular neural pathways and particular kinds of learning?

6. a. Briefly outline the neural degeneration that causes Parkinson's disease.

   b. Why can the symptoms of Parkinson's disease be viewed as motor deficits *and* failures of remembering how to do something?

7. a. Describe the patterns for predicting the weather that researchers showed both people with Parkinson's disease and unaffected people. (See Figure 14.32 in your text. Knowlton et al., 1996)

   b. Compare the results for both groups of subjects. (See Figure 14.33 in your text.)

   c. What do these results suggest about the nature of the symptoms of Parkinson's disease?

8. Let's look at some motor learning that is disrupted by damage to premotor cortex and the supplementary motor area which is shown in Figure 14.34 in your text.

   a. Describe the instrumental learning task, shown in Figure 14.35 in your text, that researchers taught normal monkeys and monkeys with lesions of the supplementary motor cortex. (Thaler et al., 1995)

   b. How well did monkeys with lesions learn the task and a task cued by an auditory stimulus?

   c. What was the nature of their deficit?

9. What is another kind of task that this kind of lesion impairs? (See Figure 14.36 in your text. Chen et al., 1995)

10. a. When monkeys are performing a memorized series of responses, how do neurons in the supplemental motor area respond? (Mushiake et al., 1991)

    b. When the task was later cued with a visual stimulus, how did response of the neurons change?

11. Which brain regions were active when human subjects learned a sequence of button presses? when they performed the task? (Hikosaka et al., 1996)

Learning Objective 14-8 Describe the role of dopamine in reinforcing brain stimulation; discuss the effects of systemic administration of dopamine antagonists and agonists.

Read pages 441-444 and answer the following questions.

1. Briefly recount the discovery of reinforcing brain stimulation by Olds and Milner (1954). (See Figure 14.37 in your text.)

2. a. The most reliable location for producing reinforcing brain stimulation is the _____

    _____ _____ which is a bundle of _____ that travels from the

    _____ to the _____ _____ _____.

    b. Where do most researchers place the tips of their electrodes?

3. a. In which midbrain structures do the major pathways of neurons whose terminal buttons secrete dopamine begin? (Lindvall, 1979; Fallon 1988)

    b. Explain why the activity of the compact clusters of the cell bodies of these neurons has a surprisingly widespread effect.

4. Complete this table describing the mesolimbic system and the mesocortical system..

Mesolimbic System	Mesocortical System
Begins in:	Begins in:
Projections	*Projections found in rats*
1.	1.
2.	2.
3.	3.
4.	*Additional projections found in primates*
5.	1.
	2.

5. a. Name and describe the location of the structure that has been the focus of much of the research on the physiology of reinforcement? (See Figure 14.38 in your text.)

    b. If dopamine receptors in the nucleus accumbens are stimulated, what kinds of behaviors are reinforced? (Routtenberg and Malsbury, 1969; Crow, 1972; Olds and Fobes, 1981; Hoebel et al., 1983; Guerin et al., 1984)

6. Describe research to determine how dopamine antagonists—drugs that block dopamine receptors— exert their effects.

    a. What task did subject rats have to perform in order to obtain reinforcing electrical stimulation? (Stellar et al., 1983)

    b. How did injections of a dopamine receptor blocker affect the reinforcing value of electrical brain stimulation? (Study Figure 14.39 in your text.)

7. a. If you do not remember the microdialysis technique, review the description in Chapter 5.

    b. What is the common effect, verified using this technique, of electrical brain stimulation, administration of cocaine or amphetamine, and natural reinforcers such as food? (Moghadam and Bunney, 1989; Nakahara et al., 1989; Phillips et al., 1992. See Figure 14.40 in your text.)

    c. What other kinds of stimuli also cause the release of dopamine in the brain? (Salamone, 1992)

Learning Objective 14-9 Discuss how the reinforcement system may detect reinforcing stimuli and strengthen synaptic connections.

Read pages 444-449 and answer the following questions.

1. Why, according to most researchers, is electrical stimulation of some regions of the brain reinforcing?

2. What two functions must be performed by the reinforcement system if a stimulus is to become reinforcing?

    1.   2.

3. Nick, having taken advantage of a special offer at the video store, watched three movies in two days. When a friend called to see if he wanted to get together and rent a movie, Nick suggested they go out for pizza instead. Use this example to explain one of the difficulties in determining what is a reinforcing event.

4. Name the brain structure that appears to be an important focal point in reinforcement.

5. Briefly describe how researchers demonstrated that dopaminergic neurons in the ventral tegmental area respond to both primary reinforcing stimuli and conditioned stimuli. (Ljungberg et al., 1992)

6. Review how neutral stimuli become conditioned reinforcers or conditioned punishers through classical conditioning.

7. List the three inputs to the ventral tegmental area that may play the most important role in reinforcement.

    1.   2.   3.

Summarize the possible roles of each of these three inputs in detection of reinforcing stimuli:

8. a. What kind of tasks are and are not disrupted by lesions of the amygdala or its disconnection from the visual system. (Spiegler and Mishkin, 1981; Gaffan et al., 1988)

    b. How did researchers teach animals to press a lever that turned on a flashing light? (Cador et al., 1989; Everitt et al., 1989)

    c. What kind of reinforcer was the flashing light?

    d. How did lesions of the amygdala affect the animals' rate of responding?

    e. What do the results suggest about the role of the amygdala in reinforcement?

9. a. How does satiety affect the firing rate of some neurons in the lateral hypothalamus and substantia innominata. (Burton et al., 1976; Rolls et al., 1986. See Figure 14.41 in your text.)

    b. What, then, may be the role of this region in reinforcement?

10. a. What neurotransmitter is secreted by the terminal buttons of axons that connect the prefrontal cortex and the ventral tegmental area?

    b. What effect does the activity of these synapses have on dopaminergic secretions in the ventral tegmental area and the nucleus accumbens in turn? (Gariano and Groves, 1988)

    c. What are some of the functions of the prefrontal cortex? (Mesulam, 1986) How may the role of the prefrontal cortex in reinforcement compliment these functions?

11. Explain the role that dopamine plays in whether synaptic strengthening takes place. (See Figure 14.42 in your text.)

12. a. What response of the neuron did Stein and Belluzzi (1989) attempt to reinforce with drug infusions?

    b. When did they successfully reinforce neural activity? (Study the graphs in Figure 14.43 in your text.)

    c. What conclusion about the effect of dopamine may be drawn from the results?

13. List some of the natural reinforcers that cause the release of dopamine in the nucleus accumbens.

14. Study Figure 14.44 in your text and explain how increasing sexual contact affected the level of extracellular dopamine in the nucleus accumbens of a male rat. (Pfaus et al., 1990)

15. a. When did the taste of saccharine cause an increase in the release of dopamine? a decrease? (Mark et al., 1989)

    b. Explain why the same stimulus had two different effects.

16. a. Why did rats press a lever to obtain food, when there already was food freely available in the operant chamber? (Salamone et al., 1991)

    b. When researchers injected a dopamine receptor blocker into the nucleus accumbens, how was their behavior affected? their hunger?

    c. In another study with food, why did all rats climb a barrier to obtain food? (Cousins et al., 1996)

    d. And how did an injection of 6-HD affect their behavior? their motor skills?

    e. What do these findings suggest is the affect of dopamine release in the nucleus accumbens on behavior reinforced by appetitive stimuli?

17. Finally, review the anatomical connections of the nucleus accumbens and then evidence on the way that dopamine released there during reinforcement affects behavior. (Scheel-Krüger and Willner, 1991)

18. a. If the prefrontal cortex of a rat is stimulated electrically or with a dopamine agonist, how is behavior affected? (Stein and Belluzzi, 1989; Duvauchelle and Ettenberg, 1991)

    b. How is behavior affected by a drug that blocks dopamine receptors there?

Lesson II Self Test

1. When an animal is trained to make a classically conditioned emotional response by pairing a tone and a foot shock, the tone is the

   a. US.

   b. CS.
   c. US.
   d. UR.

2. The synaptic changes that produce the classically conditioned emotional response occurs in the

   a. substantia nigra and the ventral tegmental area.
   b. auditory cortex and the nucleus accumbens
   c. MGm and the basolateral amygdala.
   d. medial forebrain bundle.

3. A classically conditioned response will disappear if the

   a. US is not associated with a biological need.
   b. CS is presented repeated by itself.
   c. CR was originally a behavior the animal had never made before.
   d. CR does not have favorable outcomes.

4. The principal nuclei of the basal ganglia are the

   a. caudate nucleus, the putamen, and the globus pallidus.
   b. amygdala, the substantia nigra, and the putamen.
   c. neostriatum and the substantia nigra.
   d. caudate nucleus, globus pallidus, and subthalamic nucleus.

5. People with Parkinson's disease who participated in an experiment to predict the weather from a set of cards

   a. performed as well as normal subjects.
   b. learned the task, but never improved their performance.
   c. learned the task more slowly than normal subjects.
   d. never learned the task.

6. Monkeys with lesions of the supplementary motor cortex never learned to extend their arm through an opening in the cage because they

   a. had lesion-induced difficulties with coordination.
   b. were no longer able to learn a response through instrumental conditioning.
   c. could not learn to make a self-initiated response.
   d. had lesion-induced deficits in visual perception.

7. If you wished to have the best chance of an animal pressing a lever to receive reinforcing brain stimulation, where would you place an electrode?

   a. medial forebrain bundle
   b. MGm

   c. nucleus accumbens
   d. premotor cortex

8. When rats trained to go through a runway to receive reinforcing electrical brain stimulation were given a drug that blocks dopamine receptors, the drug _____ the reinforcing effects of brain stimulation.

   a. increased
   b. replaced
   c. reduced
   d. abolished

9. The effects of reinforcing brain stimulation are _____ those of natural reinforcers.

   a. weaker than
   b. greater than
   c. similar to
   d. not as long-lasting as

10. An appetitive stimulus

   a. reinforces behavior only under certain conditions.
   b. facilitates the synthesis of acetylcholine.
   c. inhibits the release of dopamine.
   d. activates the reinforcement system only when the animal is not engaging in an appetitive behavior.

11. Infusion of dopamine or cocaine through a micropipette _____ the rate of firing of CA1 pyramidal neurons if the infusions occurred _____ spontaneous bursts of action potentials.

   a. increased; during
   b. increased; following
   c. decreased; during
   d. decreased; following

12. Injections of a dopamine blocker into the nucleus accumbens appears

   a. to make animals less motivated to perform an instrumentally conditioned response.
   b. to alter animals' ability to respond to natural reinforcers, such as food.
   c. to make animals less sensitive to biological drives.
   d. to prevent animals from recognizing familiar stimuli.

Answers for Self Tests

Lesson I			Lesson II		
1.	a	Obj. 14-1	1.	b	Obj. 14-6
2.	b	Obj. 14-1	2.	c	Obj. 14-6
3.	c	Obj. 14-1	3.	b	Obj. 14-6
4.	b	Obj. 14-2	4.	a	Obj. 14-7
5.	b	Obj. 14-2	5.	d	Obj. 14-7
6.	a	Obj. 14-2	6.	c	Obj. 14-7
7.	d	Obj. 14-3	7.	a	Obj. 14-8
8.	c	Obj. 14-3	8.	c	Obj. 14-8
9.	c	Obj. 14-3	9.	c	Obj. 14-8
10.	b	Obj. 14-4	10.	a	Obj. 14-9
11.	d	Obj. 14-4	11.	b	Obj. 14-9
12.	a	Obj. 14-5	12.	a	Obj. 14-9

CHAPTER 15
Relational Learning and Amnesia

Lesson I: Human Anterograde Amnesia

Read the interim summary on pages 464-465 of your text to re-acquaint yourself with the material in this section.

> *Learning Objective 15-1*   Describe the nature of human anterograde amnesia and the type of brain damage that causes it.

Read pages 452-456 and answer the following questions.

1. Explain the difference between anterograde amnesia and retrograde amnesia in your own words or draw a time line to illustrate the difference. (See Figure 15.1 in your text.)

2. If you are asked to describe Korsakoff's syndrome, how will you respond to the following questions?

    a. What is the most important symptom?

    b. Is speech affected?

    c. What is the usual cause of Korsakoff's syndrome and how does it lead to a vitamin deficiency?

    d. When patients with Korsakoff's syndrome are asked about recent events, what tactic do they take? What is the formal name for this behavior?

3. a. Why did patient H.M. undergo bilateral removal of the medial temporal lobe?

    b. Although the operation was successful in treating his condition, what unexpected side effect became apparent?

    c. Why did this side effect go undetected in other patients who had previously undergone this procedure?

    d. When this deficit occurs, what brain structure has been removed during surgery?

e. How has the procedure been modified to lessen the side effects and what test is administered prior to surgery as a further safeguard? (Wada and Rasmussen, 1960)

4. Briefly describe the case of patient H.M. following surgery.

   a. Why has H.M. has been studied so thoroughly?

   b. Place an *N* or an *I* in each blank to indicate whether H.M.'s ability in the following areas remains *normal* or has been *impaired* by his surgery.

   _____ intellectual ability   _____ memory of events a few years before surgery

   _____ memory for older events   _____ immediate verbal memory for numbers

   _____ memory for events after surgery   _____ immediate verbal memory for words

   _____ personality   _____ mental arithmetic computation

   c. What has H.M. said that indicates that he is aware of his condition?

   d. Under what conditions, and for how long, can H.M. remember small amounts of verbal information?

   e. Why does H.M. work well at repetitive tasks?

5. After studying H.M., what conclusions did Milner and her colleagues make about the role of the hippocampus in memory formation.

   1.

   2.

   3.

6. a. What is the capacity and duration of short-term memory? long-term memory?

   b. What is the role in memory formation of rehearsal? consolidation? (See Figure 15.2 in your text.)

   c. For patients with anterograde amnesia such as H.M., what does not appear to occur?

7. H.M. and other people with anterograde amnesia are not completely unable to learn. What categories of learning have they demonstrated under careful testing conditions?

   1.   2.   3.

8. a. Describe how Milner (1970) demonstrated that H.M. still has the ability to form perceptual memories. (See Figure 15.3 in your text.)

b. Why is the broken drawings task called a priming task?

c. Now describe the stimuli, procedure, and results of a different kind of priming task that Gabrieli and colleagues gave to H.M. and nonamnesic control subjects. (Study Figure 15.4 in your text. Gabrieli et al., 1990)

9. How did researchers demonstrate that patients with anterograde amnesia can learn to recognize melodies? faces? (Johnson et al., 1985)

10. Describe some of the stimulus-response tasks associated with sensory-response learning that amnesic subjects including H.M. have been taught. (Woodruff-Pak, 1993; Sidman et al., 1968)

11. a. Describe the mirror-drawing task that Milner (1965) used to demonstrate that H.M. is still capable of motor learning. (See Figure 15.5 in your text.)

b. How successful at mirror drawing is H.M?

c. What other motor-learning task did he learn?

Learning Objective 15-2 Discuss the distinction between declarative memories and nondeclarative memories and their relation to anterograde amnesia.

Read pages 456-460 and answer the following questions.

1. When H.M. and other amnesic subjects are asked if they remember anything about experiments they have participated in, how do they respond?

2. a. Subjects in a study of the development of episodic and emotional memories each had a particular kind of brain damage. Note it in the table below. (Bechara et al., 1995)

Subject	Region of Brain Damage	Conditioned Emotional Response	Recall of Task
S.M.			
W.C.			
R.H			

b. Describe the task and note the conditioned emotional response of each subject when the blue light went on.

c. Note the subjects' response when they were asked about the experimental procedure.

    d. Explain why the results are consistent with the kind of brain damage each subject suffered.

3. Researchers who study anterograde amnesia distinguish between two kinds of memories. Identify and define each. (Eichenbaum et al., 1992; Squire, 1992; Squire et al., 1989)

    1.

    2.

4. Classify these kinds of memories as declarative (D) or nondeclarative (ND)

    _____ catching a ball   _____ programming a VCR

    _____ swinging a golf club   _____ reminiscing at a high school reunion

    _____ listing all the places you have lived   _____ getting up on water skis

5. a. Why did Graf et al. (1984) ask amnesic and nonamnesic subjects to rate how much they liked particular six-letter words?

    b. Describe the explicit (declarative) memory and the implicit (nondeclarative) memory tests that followed and how well both groups of subjects performed. (Study Figure 15.6 in your text.)

6. a. What did Squire and his colleagues ask subjects to study? (Squire et al., 1992)

    b. How did the researchers measure regional cerebral blood flow during testing?

    c. Carefully describe the differing instructions to the subjects and the type of task each set of instructions represented.

    d. Describe regional blood flow, especially in the hippocampus, during the declarative memory task and the priming task. A PET scan of regional blood flow during testing conditions is shown in Figure 15.7 in your text.

    e. What do the results suggest about the role of the hippocampus?

To review these memory tasks study Table 15.1 in your text.

7. Anterograde amnesia appears to disrupt the ability to establish new _____ memories, but the ability to establish new _____ memories remains unchanged.

8. How did researchers demonstrate that one of the components of the anterograde amnesia of H.M. is a verbal memory deficit? (Gabrieli et al., 1988)

9. a. What do episodic memories consist of?

    b. When we want to retell them, what do we remember?

    c. What role may the hippocampal formation play in helping us to retell our experiences?

10. Why is it useful to consider the consequences of anterograde amnesia as a failure of relational learning?

Learning Objective 15-3 Review the connections of the hippocampal formation with the rest of the brain and describe evidence that damage to the hippocampal formation and related structures causes anterograde amnesia.

Read pages 460-464 and answer the following questions.

1. Complete these sentences.

    a. The hippocampal formation consists of

    b. The most important input to the hippocampal formation is

    c. The outputs of the hippocampal formation are primarily from

2. Name the region through which the hippocampal formation receives

    a. input from subcortical regions.

    b. dopaminergic input.

    c. noradrenergic input.

    d. serotonergic input

    e. acetylcholinergic input.

3. When these neurotransmitters are released, how do they affect the function of the hippocampal formation?

4. What kind of information may the mammillary bodies receive from the hippocampal formation and where is this information sent on to?

5. To review these connections study Figures 15.8 and 15.9 in your text and fill in the blanks in Figure 1 on the next page.

6. a. A temporary interruption in blood flow caused by cardiac arrest left patient R.B. with permanent anterograde amnesia. Following his death, a histological examination of his brain indicated which region of the brain was damaged? (Zola-Morgan et al., 1986)

    b. Damage to field CA1 from anoxia is shown in Figure 15.10 in your text. (Rempel-Clower et al., 1996)

    c. In addition to humans, what other species suffer anterograde amnesia when field CA1 is damaged? (Auer et al., 1989; Zola-Morgan et al., 1992)

7. a. What kind of receptors are found in abundance in field CA1?

    b. What kinds of disturbances stimulate glutamatergic terminal buttons to release high levels of glutamate?

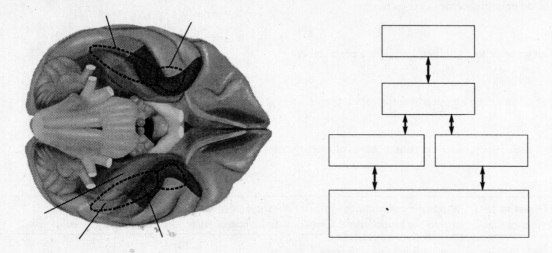

Figure 1

c. Carefully describe how a high level of glutamate affects NMDA receptors and thus explains why field CA1 is so sensitive to a lack of oxygen.

d. How can the damaging effects of a lack of oxygen be reduced? (Rothman and Olney, 1987)

8. What other structures, when damaged, contribute to amnesia? (Zola-Morgan et al., 1989b, 1993)

9. a. What brain structure is almost always severely degenerated in patients with Korsakoff's syndrome? (See Figure 15.11 in your text. Kopelman, 1995)

   b. What other brain structures, that when damaged, cause amnesia? Cite research with patients suffering from brain damage resulting from conditions other than Korsakoff's syndrome. (Calabrese et al., 1995; D'Esposito et al., 1995; McMackin et al., 1995; Malamut et al., 1992)

   c. According to most researchers, what may be the physical change that results in amnesia? What is an alternate explanation?

10. Summarize the results of a series of PET scans of a patient with Korsakoff's syndrome that suggest the physiological cause of confabulation. (Benson et al., 1996)

11. a. Describe the recognition task Schacter et al. (1996) and colleagues taught a man with right frontal lobe damage.

b. When did the man make the most "false alarms?"

c. What do these results suggest about the role of the frontal lobes?

12. When anterograde amnesia occurs without retrograde amnesia, which brain structures have been damaged? (Rempel-Clower et al., 1996; Calabrese et al., 1995; Kapur et al., 1996)

Lesson I Self Test

1. Anterograde amnesia can best be described as

   a. a failure of short-term memory.
   b. a failure to establish new nondeclarative memories.
   c. the loss of relational learning.
   d. a diminished sense of time.

2. Korsakoff's syndrome sometime results from _____that accompanies chronic alcoholism.

   a. a thiamine deficiency
   b. glucose intolerance
   c. a folic acid deficiency
   d. the pyruvate deficiency

3. Patient H.M. is

   a. aware of his disorder.
   b. bored by repetitive tasks.
   c. unable to learn any new information.
   d. frustrated by his difficulty in following a conversation.

4. When patients with anterograde amnesia are retested on an incomplete drawing task they

   a. showed a priming effect for familiar objects only.
   b. showed a priming effect for geometric patterns but not for real objects.
   c. showed a priming effect for all stimuli.
   d. did not show any signs of a priming effect.

5. Patients who showed a conditioned emotional response to a blue light that preceded a loud boat horn had bilateral damage to the

   a. hippocampus.
   b. amygdala.
   c. mammillary bodies.
   d. thalamus.

6. Nondeclarative memories

   a. are a form of perceptual memory.
   b. are usually expressed in writing.
   c. fade more quickly than declarative memories.
   d. do not require deliberate efforts to learn information.

7. What we remember about an episode that permits us to store it in long-term memory

   a. are a few striking details.
   b. are the events and the context in which they occur.
   c. is the information that is inconsistent with what we already know.
   d. is the information about it that we rehearse.

8. You looked up the telephone number, but went to let the dog in before making the call and found you had forgotten the number. This occurred because

   a. of the phenomenon of retrograde amnesia.
   b. long-term memories are difficult to retrieve.
   c. the phone number exceeded the capacity of short-term memory.
   d. insufficient rehearsal did not enable short-term memory to be converted into long-term memory.

9. The most important input to the hippocampal formation is the

   a. anterior thalamus.
   b. locus coeruleus
   c. entorhinal cortex.
   d. subiculum.

10. Neurons in field CA1 of the hippocampus are easily damaged when metabolic disturbances set off a series of events that include the

a. rapid proliferation of NMDA receptors.

b. desynchronized firing of presynaptic axons.

c. entry of calcium into the neurons resulting in excessively high levels of intracellular calcium.

d. release serotonin at abnormally high levels by serotonergic terminal buttons.

11. Confabulation, a symptom of Korsakoff's syndrome, may be caused by damage to the

a. prefrontal cortex.

b. mammillary bodies.

c. perirhinal cortex

d. amygdala.

12. During testing, a patient made many more "false alarms" when the

a. researcher distracted him.

b. incorrect stimulus resembled one he had previously seen or heard.

c. length of time between training and testing increased.

d. number of choices increased.

Lesson II: Relational Learning in Laboratory Animals

Read the interim summary on pages 475-476 to re-acquaint yourself with the material in this section.

Learning Objective 15-4   Describe the role of the hippocampus in relational learning including spatial learning.

Read pages 465-467 and answer the following questions.

1. a. Describe the radial maze shown in Figure 15.12 in your text.

    b. Now describe the task used by Olton and Samuelson (1976) and the performance of the rats.

    c. In later research, how did normal rats perform if they were prevented from visiting the arms in a particular order? (Olton et al., 1977)

    d. What abilities, tested by the radial maze task, must rats have in order to survive?

    e. What kind of lesions severely disrupt the ability to avoid revisiting a place where food was just found. (reviewed by Olton, 1983)

    f. Describe how Olton and Papas (1979) used a different radial maze, shown in Figure 15.13, to test the implicit and explicit learning skills of subject rats.

    g. What kind of lesions disrupted performance and how was it disrupted?

    h. Olton (1983) suggests that lesions of the _____ or its _____ damage _____ memory, but leave _____ memory relatively intact.

2. Explain why the rats' performance in both mazes can also be explained in terms of relational memory.

3. Describe research by Morris et al., (1982) on the effect of hippocampal lesions on spatial abilities.

    a. Describe the "milk maze" the researchers used.

    b. How were rat subjects trained in the maze?

    c. What was the only kind of information rats could use to find the submerged platform?

    d. How successfully did normal rats, control rats with neocortical lesions, and rats with hippocampal lesions maneuver through the maze? (See Figure 15.14 in your text.)

    e. What do the results suggest about the importance of the hippocampal formation in relational learning?

4. a. How can the milk maze be modified to test stimulus-response learning, which is nonrelational?

    b. How easily do rats with hippocampal lesions now find the platform?

5. Summarize research on other species that confirms the importance of the hippocampus in spatial learning.

    a. How do hippocampal lesions affect the ability of homing pigeons to get their initial bearing? to keep tract of where they are near the end of their flight? (Bingman and Mench, 1990)

    b. Compare the size of the hippocampus of

        1. homing pigeons with other breeds of pigeons lacking good navigational abilities. (Rehkämper et al., 1988)

        2. birds and rodents that store food with those who do not. (Sherry et al., 1992)

        3. black-capped chickadees throughout the year. (DeVoogd, 1995)

Learning Objective 15-5   Discuss the function of place cells in the hippocampal formation.

Read pages 467-470 and answer the following questions.

1. a. In general, how did single pyramidal cells in the hippocampus respond as rats moved around their environment? (Be sure to mention spatial receptive fields in your answer. O'Keefe and Dostrovsky, 1971)

    b. What were these neurons named?

2. a. Describe how hippocampal place cells respond to environmental cues as a rat navigates the maze shown in Figure 15.15 in your text.

    b. If the maze is rotated 90 degrees, how do the place cells respond?

    c. In a symmetrical chamber, what kind of cues do rats use to orient themselves?

    d. How do place cells respond when these cues are moved as a group? interchanged?

3. Describe research by Hill and Best (1981) on how rats determine location using internally generated stimuli.

    a. How were subject rats prevented from using external cues to determine location?

    b. How did the spatial receptive fields of rats respond when they were

        1. placed in a maze that was rotated?

        2. first spun around and then placed in the maze?

4. Outline a different procedure leading to the same results used by McNaughton et al., (1989).

5. Discuss evidence on how the rat keeps track of its location in the absence of external cues obtained by recording from

    a. single neurons in the parietal cortex while a rat moved about in a maze. (McNaughton et al., 1989)

    b. hippocampal place cells while thirsty rats moved about in a square chamber which was rotated. (Wiener et al., 1995)

6. a. The hippocampus appears to receive its spatial information through the _____ _____.

    b. What property do neurons in entorhinal cortex, hippocampal pyramidal cells, and single granule cells in the dentate gyrus all share to varying degrees? (Quirk et al., 1992; Rose, 1993)

7. a. Why is it incorrect to conclude that each neuron in a spatial receptive field encodes a particular location? (Muller and Kubie, 1987)

    b. According to most researchers, how do rats learn their way through a new environment?

    c. Cite research on the stability of these "maps." (Thompson and Best, 1990)

8. a. Describe another maze, this time constructed by Young et al., (1994).

    b. Figure 15.16 in your text shows the response of a single hippocampal neuron as the rat visited particular locations in the maze. To what kind of cues did some place cells respond? did most place cells?

9. a. Although some neurons in the hippocampus of monkeys respond to location, most of them respond to a different cue. Name it.

    b. What name has Rolls (1996) given these neurons and why does he think they are present.

Learning Objective 15-6 Describe how changes in synaptic strength and monoaminergic and acetylcholinergic input may affect hippocampal functioning.

Read pages 470-472 and answer the following questions.

1. What physical changes that may have resulted from experiences involving the hippocampal formation, were observed in the brains of

    a. rats who had been raised in a complex environment. (Green and Greenough, 1986)

    b. weanling mice who had been put in a complex environment for 40 days. (Kemperman et al., 1997)

    c. rats who learned a radial-arm maze? (Mitsuno et al., 1994)

    d. animals who took part in spatial learning tasks? (Van der Zee et al., 1992; Tan and Liang, 1996)

2. Summarize the results of research using procedures that interfere with long-term potentiation and affect learning. How do targeted mutations of

    a. genes responsible for the production of PKC and CaMKII affect long-term potentiation? the ability to learn the Morris milk maze? (Grant et al., 1992; Silva et al., 1992b)

b. the NMDA receptor gene affecting only CA1 pyramidal cells affect receptor development? (See Figure 15.17 in your text. McHugh et al., 1996; Tsien et al., 1996)

3. How did the lack of NMDA receptors in the CA1 pyramidal cells affect long-term potentiation and the ability to learn the Morris milk maze?

4. In general, how may input from acetylcholinergic, noradrenergic, dopaminergic, and serotonergic neurons affect the information processing functions of the hippocampal formation?

5. Describe the effects of these neurotransmitters on the establishment of long-term potentiation.

   a. serotonin (For example, Stäubli and Otaky, 1994)

   b. norepinephrine (For example, Dahl and Sarvey, 1989)

   c. dopamine (Stein and Belluzzi, 1989)

6. Define hippocampal *theta waves* in your own words, mentioning their source and their role in long-term potentiation.

7. What happens if

   a. bursts of electrical stimulation coincide with the peaks of theta activity? (Pavlides et al., 1988)

   b. depolarizing stimulation coincides with the peaks of theta activity? with the troughs? (See Figure 15.18 in your text. Huerta and Lisman, 1996)

   c. animals receive injections of scopolamine, a drug that blocks muscarinic acetylcholine receptors? (Givens and Olton, 1990)

   d. animals receive low doses of alcohol? (Givens, 1995)

8. a. What procedure appears to reduce the effects of damage to cholinergic inputs to the hippocampal formation?

   b. How do these transplants affect

      1. theta activity and spatial receptive fields? (Buzsáki et al., 1987; Shapiro et al., 1989)

2. performance in the Morris milk maze? (Nilsson et al., 1987; Ridley et al., 1991, 1992)

3. relational learning in older animals? (Gage et al., 1984; Olton et al., 1991)

9. Give several examples of theta behaviors and non-theta behaviors.

10. a. What is the relationship between the rate of a rat's sniffing and the waves of hippocampal theta rhythm? (Wiener et al., 1989)

 b. What may happen to environmental information the rat is sniffing out when theta waves are present, when they cease, and during slow-wave sleep?

Learning Objective 15-7 Outline a possible explanation of the role of the hippocampal formation in learning and memory.

Read pages 472-475 and answer the following questions.

1. What have many investigators concluded about the deficit in spatial learning caused by hippocampal lesions. (For example, Wiener et al., 1989; Sutherland and Rudy, 1989)

2. What may have been the original function of the hippocampus? later functions?

3. Draw on your own experiences to explain how moving from place to place depends on recognizing the context of the stimuli you encounter and their relationship to each other.

4. a. Let's look at time as a contextual stimulus. Describe the maze and the nonspatial alternation task Raffaele and Olton (1988) taught rats.

 b. What information did rats need to remember to succeed?

 c. How well do normal rats learn this task? rats with lesions of the fornix?

 d. What do the results suggest about the function of the hippocampus?

5. Review the inputs to the hippocampal formation that may explain how events are placed in their proper context.

6. a. Describe how Phillips and LeDoux (1992) established a conditioned emotional response in rats to a context and to a specific stimulus.

   b. During testing, how did each group of rats respond? (See Figure 15.19 in your text.)

      1. control rats

      2. rats with lesions of the amygdala

      3. rats with lesions of the hippocampus

   c. What do the results suggest about the role of the hippocampus in distinguishing particular contexts?

7. a. The hippocampal formation is connected directly to the basolateral amygdala by means of the

      _____ _____ _____.

   b. How can long-term potentiation be established in the basolateral amygdala? blocked? (Maren and Fanselow, 1995)

   c. How can a conditioned emotional response to a contextual stimuli be disrupted?

8. Now refer to the learning experiences of a normal person and patients like H.M. to explain how anterograde amnesia might be an inability to distinguish and relate different contextual stimuli to each other.

9. Outline the hypothetical model concerning the role of the hippocampal formation in learning and memory proposed by Rolls (1989; 1996)

   a. To which region does the neocortex send information about events and episodes?

   b. And where does this region then send the information which has now been analyzed further?

   c. Trace the circuit that ends with the information being returned to the neocortex.

   d. What kind of axons are found in field CA3?

   e. Explain how the network of neurons in field CA3 might function as an autoassociator.

10. What capability of the hippocampal formation may permits us to recall our actions with accuracy?

Lesson II Self Test

1. Rats with hippocampal lesions could not efficiently visit the arms of a radial maze because they could not

   a. distinguish between the many arms of the maze.
   b. learn which arms never contained food.
   c. establish an initial bearing.
   d. remember where they had just been.

2. Information in reference memory is

   a. constantly being replaced.
   b. more difficult to retrieve than information in working memory.
   c. relatively permanent.
   d. episodic in nature.

3. Rats are trained in a milk maze

   a. to reduce the effects of tactile stimulation.
   b. to test their spatial perception and memory.
   c. to avoid using food as a reinforcing stimulus.
   d. to assess their stimulus-response learning.

4. How do rats in a symmetrical chamber react when researchers move environmental stimuli as a group?

   a. Rats move toward the center of the chamber.
   b. Rats tend to remain in one location.
   c. Rats run constantly around the perimeter of the chamber.
   d. Rats reorient their responses accordingly.

5. The hippocampus appears to receive its spatial information through the

   a. fornix.
   b. entorhinal cortex.
   c. amygdala.
   d. medial septum.

6. Place cells in primates tend to respond best to

   a. where the animal is located.
   b. where the animal is looking.
   c. the animal's internally generated stimuli.
   d. the direction of the animal's locomotion.

7. What is the source of theta rhythms?

   a. glutamatergic axons from the dentate gyrus

   b. serotonergic axons from the fornix
   c. dopaminergic axons from the substantia nigra
   d. acetylcholinergic axons from the medial septum

8. Theta behaviors closely associated with hippocampal theta activity include

   a. ingestive behaviors.
   b. wakefulness and sleep
   c. reproductive behaviors.
   d. exploration or investigation.

9. Depolarizing stimulation that coincided with the peaks of theta waves resulted in _____.

   a. depression
   b. disorientation of hippocampal place cells
   c. long-term potentiation
   d. disruption of working memory

10. The original function of the hippocampus may have been

   a. to regulate an animal's circadian rhythms.
   b. to help an animal recognize new stimuli.
   c. to regulate an animal's metabolism in response to environmental changes.
   d. to help an animal navigate in its environment.

11. Stimulation of the ventral angular bundle _____ long-term potentiation in the basolateral amygdala.

   a. produced.
   b. abolished.
   c. improved
   d. had no effect on

12. Rolls suggests that neurons in field CA3 function as an autoassociator—that is, they

   a. can produce the appropriate output from fragments of the original pattern.
   b. link together the structures that make up the hippocampal formation.
   c. along with the recurrent collaterals provide an alternate system of connections that can minimize the effects of brain damage.
   d. somehow connecting the hippocampal formation with association cortex.

Answers for Self Tests

Lesson I		
1.	c	Obj. 15-1
2.	a	Obj. 15-1
3.	a	Obj. 15-1
4.	c	Obj. 15-1
5.	a	Obj. 15-2
6.	d	Obj. 15-2
7.	b	Obj. 15-2
8.	d	Obj. 15-2
9.	c	Obj. 15-3
10.	c	Obj. 15-3
11.	a	Obj. 15-3
12.	b	Obj. 15-3

Lesson II		
1.	d	Obj. 15-4
2.	c	Obj. 15-4
3.	b	Obj. 15-4
4.	d	Obj. 15-5
5.	b	Obj. 15-5
6.	b	Obj. 15-5
7.	d	Obj. 15-6
8.	d	Obj. 15-6
9.	c	Obj. 15-6
10.	d	Obj. 15-7
11.	a	Obj. 15-7
12.	a	Obj. 15-7

CHAPTER 16
Human Communication

Lesson I: Brain Mechanisms of Speech and Comprehension

Read the interim summary on page 495 of your text to re-acquaint yourself with the material in this section.

> *Learning Objective 16-1* Describe the use of subjects with brain damage in the study of language and explain the concept of lateralization.

Read pages 478-479 and answer the following questions.

1. List three medical conditions that researchers often study to learn more about the effects that physical damage to the brain has on speech.

    1.   3.

    2.

2. Underline the most frequently studied of these conditions.

3. Name and describe the most important category of speech disorders.

4. Explain this sentence: "Verbal behavior is a lateralized function."

5. a. Describe the Wada test and explain how it is used to determine hemispheric dominance for speech.

    b. Which hemisphere most often controls speech?

6. a. Explain why the left hemisphere is particularly suited for speech.

    b. Explain how the right hemisphere contributes to speech. (Gardner et al., 1983)

Learning Objective 16-2 Describe Broca's aphasia and the three major speech deficits that result from damage to Broca's area: agrammatism, anomia, and articulation difficulties.

Read pages 479-484 and answer the following questions.

1. Describe how Broca's aphasia affects

    a. speech production.

    b. grammar, especially the use of function words and content words.

    c. comprehension.

2. a. Broca's aphasia, named for Paul Broca, appears to result from damage to what region of the brain? (See Figure 16.1 in your text.)

    b. Describe more precisely the areas in the vicinity of Broca's area that, when damaged, do and do not produce Broca's aphasia. (H. Damasio, 1989; Naeser et al., 1989)

    c. What other region if damaged produces a Broca-like aphasia? (Study Figure 16.2 in your text. Damasio et al., 1984; Leblanc et al., 1992)

3. What did Wernicke (1874) suggest is the speech function of Broca's area?

4. a. Name and briefly describe three deficits that are characteristic of Broca's aphasia, suggesting that the disorder is not a simple one.

    1.

    2.

    3

b. Which of these deficits is the most elementary? the most complex?

5. a. Describe the task and the cards that Schwartz and her colleagues (1980) used to test the speech comprehension skills of people with Broca's aphasia. (See Figure 16.3 in your text.)

 b. How often did the subjects select the appropriate picture?

 c. What do these results suggest about the ability of Broca's aphasics to use grammatical information?

6. a. What critical location for control of speech articulation did Dronkers (1996) find? (See Figure 16.4 in your text.)

 b. How did she locate it? Be sure to mention apraxia of speech in your answer. (See Figure 16.5 in your text.)

7. a. The agrammatism and anomia characteristic of Broca's aphasia are normally caused by damage to what area of the brain?

 b. Describe how Stromswold et al. (1996) confirmed this location.

8. a. How well did Broca's aphasics respond to a sequence of verbal commands? (Boller and Dennis, 1979)

 b. What do these results suggest about the functions of the left frontal lobe?

9. How does damage to the cerebellum affect speech production? (Adams and Victor, 1991) grammar? (Silveri et al., 1994).

Learning Objective 16-3 Describe the symptoms of Wernicke's aphasia, pure word deafness, and transcortical sensory aphasia and explain how they are related.

Read pages 484-487 and answer the following questions.

1. Return to Figure 16.1 and describe the location of Wernicke's area. What is its function?

2. List the two primary characteristics of Wernicke's aphasia.

 1.   2.

3. a. Compare the fluency, articulation, inflection, and grammar of the speech of patients with Broca's aphasia and Wernicke's aphasia.

b. How is the speech comprehension of people with Wernicke's aphasia tested? Why?

4. Describe the reaction of patients with Wernicke's aphasia to their own speech difficulties.

5. What kind of information may be stored in Wernicke's area?

6. List three characteristic deficits resulting from damage to Wernicke's area.

    1.   3.

    2.

7. Explain the distinction between recognizing and comprehending a word.

8. a. Describe how these abilities are affected in cases of pure word deafness.

    1. comprehension of speech

    2. comprehension of nonspeech sounds

    3. comprehension of emotion expressed through intonation

    4. speech production

    b. What methods do people with this disorder use to understand and communicate with other people?

9. Describe the different functions of the left and right hemispheres in the analysis of speech sounds.

10. Outline the hypothesis of Phillips and Farmer (1990) concerning the recognition of acoustical events of short duration.

11. Study Figure 16.6 in your text and discuss the two types of brain damage that cause pure word deafness.

12. a. Now study Figure 16.7 in your text and describe the location and the presumed function of the posterior language area. Be sure to note its proximity to Wernicke's area.

b. Damage to the posterior language area results in the disorder _____ _____

_____.

13. a. Describe how these abilities are affected in cases of transcortical sensory aphasia.

1. speech comprehension

2. speech recognition and repetition

3. production of meaningful speech

b. How do the deficits of people with transcortical sensory aphasia compare with those of people with Wernicke's aphasia?

14. Because patients can repeat what they hear, what kind of connection must exist within the brain? (Return to Figure 16.7.)

15. a. Describe how extensive brain damage resulting from carbon monoxide affected the patient's speech recognition, comprehension and production. (Geschwind et al., 1968)

b. What does this case confirm about the brain mechanisms of speech?

Learning Objective 16-4 Discuss the brain mechanisms that underlie our ability to understand the meaning of words and to express our own thoughts and perceptions in words.

Read pages 487-489 and answer the following questions.

1. Words have meaning for us because they evoke particular _____, which are not stored in primary _____ _____, but in other parts of the brain.

2. Study Figure 16.8 in your text and follow the pathway responsible for recognizing and comprehending a spoken word.

3. a. Following a stroke that damaged part of the right parietal lobe, which plays a role in spatial perception, what kind of difficulty did a patient have in describing spatial relationships?

b. She could understand some meanings of particular words, but not other meanings. Explain and give examples.

4. Name and describe the specific comprehension difficulties associated with

a. damage to the association cortex of the left parietal lobe. Be sure to use the term *autotopagnosia* in your answer.

b. damage to the left temporal lobe. (McCarthy and Warrington, 1988)

c. widespread damage to the temporal and parietal lobes. (Damasio and Tranel, 1990; Hodges et al., 1992)

5. What hemisphere is involved in the comprehension of abstract aspects of speech? How did researchers determine this? (Brownell et al., 1983,1990; Bottini et al., 1994; Nichelli et al., 1995)

Learning Objective 16-5   Describe the symptoms of conduction aphasia and anomic aphasia, including aphasia in deaf people.

Read pages 489-495 and answer the following questions.

1. a. The _____ _____ is a direct connection between Wernicke's area and Broca's area.

    b. What kind of information does it presumably convey?

2. a. Describe the brain damage that causes conduction aphasia (Study Figure 16.9 in your text. Damasio and Damasio, 1980)

    b. Now describe its symptoms. Refer to specific examples from patients (Margolin and Walker, 1981).

3. Study Figure 16.10 in your text and carefully describe the direct and indirect pathways that may connect the speech mechanisms of the temporal and frontal lobes suggested by the communication deficits of patients with transcortical sensory aphasia or conduction aphasia.

4. To review what you have learned, draw in and label the following regions in Figure 1 on the next page: Broca's area, Wernicke's area, posterior language area, primary auditory cortex, the arcuate fasciculus, and the location of perceptions and memories. Draw in and label arrows responsible for the repetition of a perceived word and the translation of thoughts into words. The information needed to complete the figure is found in Figures 16.7-16.10 in your text.

5. a. The symptoms of conduction aphasia suggest that the connection between Wernicke's area and Broca's area plays what role?

    b. Describe the circuit that Baddeley (1992) refers to as the phonological loop.

    c. Summarize research that supports this hypothesis. (See Figure 16.11 in your text. Paulesu et al., 1993; Fiez et al., 1996)

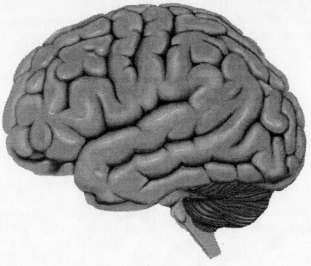

Figure 1

6. Read over the description of the picture in Figure 16.12 in your text given by the woman with anomic aphasia reported by Margolin and his colleagues. (Margolin et al., 1985)

a. How did anomic aphasia affect her speech production and comprehension?

b. When did she use circumlocutions?

c. Study Figure 16.13 in your text and describe the brain damaged sustained by this patient.

d. In follow-up informal testing, what word-finding difficulty became more apparent?

7. Begin a review of the range of verbal deficit patterns experienced by people with anomia.

a. Describe the deficits experienced by people studied by

1   Manning and Campbell (1992).

2. Semenza and Zettin (1989).

b. According to Damasio and colleagues (Damasio et al., 1991), which region has been damaged when patients have anomia for proper nouns? anomia for common nouns? (See Figure 16.14 in your text.)

c. What do they suggest is an important distinction between the two types of words?

d. What, then, may be a function of the cortex of the temporal pole?

e. Anomia for verbs results from damage to the frontal cortex. Explain why this finding is consistent with our knowledge of the functions of the frontal lobes.

f. Study Figure 16.15 in your text and briefly describe research that confirms the importance of Broca's area and the surrounding region in the production of verbs. (Petersen et al., 1988; Wise et al., 1991; McCarthy et al, 1993; Fiez et al., 1996)

8. To review: Describe the probable flow of information through the brain for the comprehension of speech and production of spontaneous speech.

9. In what hemisphere might we expect to find damage in deaf people with aphasia? Why? Where are the lesions of aphasic deaf people? (Hickok et al., 1996)

10. Identify the parts of the brain that are activated when lipreading. (See Figure 16.16 in your text. Campbell et al., 1996; Calvert et al., 1997)

11. Explain prosody in your own words.

12. How do we indicate some of the elements of prosody when we write?

13. Contrast the prosody of the speech of people with Wernicke's aphasia, fluent aphasia, and Broca's aphasia.

14. a. Which hemisphere appears to mediate prosody?

b. Name some of the other functions of this hemisphere that are probably related to prosody.

15. a. Describe the experimental tasks Weintraub et al. (1981) used to test the use and recognition of prosody by subjects with right-hemisphere damage.

b. How well did subjects perform in these experiments?

16. To review your understanding of the deficits associated with each of the disorders covered in this lesson, fill in the blanks in the table below. Compare your answers with the information in Table 16.1 in your text.

Disorder	Area of lesion	Spontaneous speech	Comprehension	Repetition	Naming
Wernicke's aphasia					
Pure word deafness					
Broca's aphasia					
Conduction aphasia					
Anomic aphasia					
Transcortical sensory aphasia					

Lesson I Self Test

1. Most observations on the physiology of language have been made through work with subjects who have sustained

   a. head injuries.
   b. cerebrovascular accidents.
   c. brain tumors.
   d. infections.

2. When we say that verbal behavior is a lateralized function, we mean that

   a. both hemispheres of the brain are equally important for speech.
   b. both hemispheres of the brain have the capacity to perform all aspects of speech.
   c. one hemisphere is dominant for speech and the other hemisphere plays no role.
   d. one hemisphere is dominant for speech and the other hemisphere plays a smaller role.

3. People with Broca's aphasia

   a. speak fluently, but have anomia.
   b. speak fluently, but have poor comprehension.

   c. speak slowly, with difficulty, but grammatically.
   d. speak slowly, with difficulty, but with meaning.

4. Anomia is a difficulty in

   a. finding the correct word to describe an object, action, or situation.
   b. pronouncing abstract, but not concrete, words.
   c. repeating sequences longer than three words.
   d. using function and content words.

5. People with Wernicke's aphasia

   a. speak fluently but without meaning and have poor comprehension.
   b. speak haltingly but with meaning and have excellent comprehension.
   c. can no longer speak but still comprehend the speech of others.
   d. speak sporadically but to the point.

6. Research with people who have transcortical sensory aphasia suggests that

a. the ability to understand is necessary for accurate memorization.

b. recognition, repetition, and rhythm are inseparable aspects of speech production.

c. the ability to speak spontaneously is necessary in order to repeat what is heard.

d. brain mechanisms needed for recognition and comprehension of speech are different.

7. Pure word deafness is the inability

a. to comprehend the meaning of words.
b. to understand the speech of others.
c. to communicate orally with others.
d. to use grammatical constructions.

8. Damage to the posterior language area disrupts the ability to

a. recognize written or spoken words.
b. learn new written or spoken words.
c. understand words and produce meaningful speech.
d. understand particular categories of words.

9. Which is not true of the arcuate fasciculus?

a. It conveys meanings of words but not sounds.
b. It conveys sounds of words but not meanings.
c. It is an arch-shaped bundle that connects Wernicke's area and Broca's area.

d. Damage results in conduction aphasia.

10. Researchers suggest that people with conduction aphasia can repeat words only if they

a. are spelled regularly.
b. have meaning.
c. have no threatening emotional content.
d. are short.

11. Aphasia in deaf people

a. results in an inability to lipread, and is caused by damage to the right hemisphere, which is involved in recognizing faces.

b. is hard to determine, because sign language is not a full-fledged language.

c. is caused by right hemisphere damage, because sign language has a visual, spatial nature.

d. is caused by left hemisphere damage, as it is in hearing people.

12. Which of the following disorders is characterized by poor prosody?

a. Wernicke's aphasia
b. Broca's aphasia
c. anomic aphasia
d. conduction aphasia

Lesson II: Reading and Writing Disorders

Read the interim summary on page 507 of your text to re-acquaint yourself with the material in this section.

Learning Objective 16-6 Describe pure alexia and explain why this disorder is caused by damage to two specific parts of the brain.

Read pages 496-499 and answer the following questions.

1. Make a general statement about the relationship between the speaking and understanding skills and the reading and writing skills of people with aphasia.

2. To illustrate this general relationship, review the speaking , understanding, and reading skills of people with the following aphasias which are summarized in tables is both your text and the study guide.

a. Wernicke's aphasia

b. Broca's aphasia

c. conduction aphasia

d. transcortical sensory aphasia

3. Briefly describe the abilities of a patient with severe fluent aphasia studied by Semenza et al. (1992) and explain what these rare exceptions suggest about neural organization.

4. Describe how pure alexia affects

   a. reading.

   b. writing. (For an interesting example, see Figure 16.17 in your text.)

   c. recognition of orally spelled words.

5. a. Study Figure 16.18 in your text and describe the lesions that cause pure alexia. Describe the neural circuits that are disrupted, paying special attention to the role of the corpus callosum. (Damasio and Damasio, 1983, 1986)

   b. If the model presented in Figure 16.18 is correct, what would we predict a lesion restricted to the posterior corpus callosum would cause? Cite research to support your answer. (Binder et al., 1992)

6. Describe the disorder in deaf people that is similar to pure alexia. What brain damage causes this disorder?

7. If people with pure alexia can recognize objects but cannot read and people with visual agnosia can read but cannot recognize objects, what does that suggest about the brain mechanisms for reading and object and word recognition and their evolution?

8. a. Which of the four types of visual stimuli that patients saw while their regional cerebral blood flow was measured by a PET scanner activated the extrastriate cortex? which did not? (See Figure 16.19 in your text. Petersen et al., 1990)

    b. What do these results suggest about its function?

Learning Objective 16-7 Describe whole-word and phonetic reading and discuss five categories of acquired dyslexias.

Read pages 499-503 and answer the following questions.

1. a. Explain the distinction between whole-word reading and phonetic reading.

    b. Under what circumstances do we use the whole-word and phonetic methods?

    c. Look at Figure 16.20 in your text for an illustration of some elements of the reading process.

2. Contrast acquired dyslexia with developmental dyslexia.

3. Let's examine some of the reading disorders resulting from acquired dyslexia, beginning with surface dyslexia.

    a. What method do people with surface dyslexia use to read? (Study Figure 16.21 in your text.)

    b. How does the spelling of a word or nonword affect their reading ability?

    c. What is the opposite reading method that people with phonological dyslexia use to read? (Study Figure 16.22 in your text.)

    d. How does phonological dyslexia affect the ability to read familiar words, unfamiliar words, and pronounceable nonwords? (Beauvois and Dérouesné, 1979; Dérouesné and Beauvois, 1979)

4. Explain why the reading ability of people with phonological dyslexia is further evidence for a distinction between whole word reading and phonetic reading.

5. Continue to discuss this distinction by describing the way words expressed in Japanese kanji and kana symbols are read.

6. What have studies of Japanese people with localized brain damage indicated about the reading process? (Iwata, 1984; Sakurai et al., 1994))

7. What reading methods appears to be lost in word-form or spelling dyslexia? (See Figure 16.23 in your text.)

8. Explain how people with word-form dyslexia read by describing their ability to recognize words if

    a. permitted to spell the word.

    b. someone else spells the word aloud.

    c. they spell the word incorrectly due to the severity of their deficit.

9. Explain why direct dyslexia resembles transcortical sensory aphasia. (Schwartz et al., 1979; Lytton and Brust, 1989)

10. Describe how direct dyslexia affects the ability

    a. to read aloud, and to understand what is read.

    b. to communicate with others, and to understand what is said.

11. Describe how brain injury resulting from an automobile accident affected patient R.F.'s ability to

    a. read most words.

    b. name most common objects.

    c. communicate with others.

    d. choose words that matched a picture. (See Figure 16.24 in your text. Margolin et al., 1985)

12. a. What reading method appears to be lost? is still partially intact?

   b. What observation about spelling did R.F. make that confirms this assessment?

Learning Objective 16-8   Explain the relation between speaking and writing and describe the symptoms of phonological dysgraphia, orthographic dysgraphia and semantic (direct) dysgraphia.

Read pages 503-504 and answer the following questions.

1. Explain why people with speech difficulties may also have writing difficulties.

2. Describe some of the remarkably specific writing deficits related to difficulties in motor control. (Cubelli, 1991; Alexander et al., 1992; Margolin and Goodman-Schulman, 1992; Silvers, 1996)

3. Let's look at four ways of writing words. If you have not already tried the antidisestablishmentarianism experiment, try it now. Use you own experience to describe any difficulties that you had when you wrote the word

   a. without saying it softly to yourself.

   b. and softly sang a song at the same time.

4. Continue by describing three other writing methods. Be sure to note what type of sensory information is required.

   1.

   2.

   3.

5. Let's examine research on phonetic and visual spelling. Describe how phonological dysgraphia affects the ability to write

   a. nonsense words by sounding them out.

   b. familiar and unfamiliar words by imagining how they look.

6. Describe how orthographic dysgraphia affects the ability to write

   a. regular words and nonsense words.

   b. irregular words. (Beauvois and Dérouesné, 1981)

7. Phonological dysgraphia then is impaired _____ writing resulting from damage to the _____ _____ lobe and orthographic dysgraphia is impaired _____ writing resulting from damage to the _____ _____ lobe. (Benson and Geschwind, 1985)

8. Compare the writing deficits of Japanese patients with those of patients whose languages use the Roman alphabet. (Iwata, 1984; Yokota et al., 1990)

9. a. See Figure 16.25 in your text and then describe the writing skills of the Japanese man studied by Kawamura et al., (1989).

   b. What do these results suggest about the organization of writing in the brain? (Study Figure 16.26 in your text.)

10. What is the characteristic symptom of semantic agraphia? (Roeltgen et al, 1986; Lesser, 1989)

Learning Objective 16-9 Describe research on the neurological basis of developmental dyslexias.

Read pages 504-507 and answer the following questions.

1. Describe evidence that suggests there is a biological component of developmental dyslexia. (Pennington et al., 1991; Wolff and Melngailis, 1994; Grigorenko et al., 1997)

2. Describe the nature of the deficit shown by most people with developmental dyslexia. (Castles and Coltheart, 1993)

3. Summarize research in which the brains of deceased people who had developmental dyslexia were studied. (Galaburda et al., 1985; Galaburda, 1988; Humphreys et al, 1990)

    a. In all cases, which brain region appeared abnormal?

    b. Study Figure 16.27 in your text and compare the arrangement of cells in the left planum temporale of a normal person and a person with developmental dyslexia.

    c. What may be the cause of this irregular cell arrangement?

4. What did Filipek's (1995) literature search reveal about the location of brain damage that causes developmental dyslexia?

5. Galaburda and Livingstone's (1993) research on dyslexia received considerable attention. Let's review their finding and implications.

    a. Where did they find a deficit in the brains of dyslexics?

    b. Specifically, what was wrong with this part of their brains?

    c. Why might this damage result in dyslexia? Explain Stein and Walsh's (1997) hypothesis.

    d. Describe further supporting research. (See Figure 16.28 in your text. Cornelissen et al., 1991; Walsh and Butler, 1996; Eden et al., 1996)

6. When Geschwind and Behan (1984) compared the reading ability and incidence of immune diseases of a group of right- and left-handed people, what did they learn?

7. Explain Geschwind and Behan's suggestion that left-handedness and developmental dyslexia interact.

8. What specific mechanism do Humphreys and colleagues suspect causes developmental dyslexia? (Humphreys et al., 1990)

Lesson II Self Test

1. Pure alexia is a(n) _____ disorder and patients _____.

    a. perceptual; can no longer read, but can still write
    b. motor; can no longer read or write
    c. sensory; can read but no longer write
    d. auditory; can read and write silently but cannot read or spell aloud

2. Research using a PET scanner to measure regional cerebral blood flow suggests that a region of extrastriate cortex is important for

    a. recognition of familiar combinations of letters.
    b. encoding acoustical information about the sounds of letters.
    c. associating words with their meanings.
    d. reading speed.

3. Surface dyslexia involves a deficit in

    a. phonetic reading.
    b. whole-word reading.
    c. letter-to-sound decoding.
    d. comprehension only; reading is intact.

4. People with phonological dyslexia will have difficulty reading

    a. function words.
    b. abstract words.
    c. nonwords.
    d. content words.

5. In which form of dyslexia do people fail to comprehend what they read?

    a. word-form dyslexia
    b. phonological dyslexia
    c. direct dyslexia
    d. surface dyslexia

6. Studies of Japanese people with localized brain damage who have difficulty reading kana or kanji symbols provide evidence

    a. for reading forms based on the type of alphabet used in the language.
    b. for a universal reading form that involves brain mechanisms that existed before the invention of writing.

    c. that the brain contains redundant neural circuits involved in reading.
    d. for two different forms of reading that involve different brain mechanisms.

7. The fact that it is more difficult to write the word *antidisestablishmentarianism* while singing suggests that

    a. the "auditory image" of words expressed in music is stronger than the image of words expressed in speech.
    b. the ability to write some words depends on being able to articulate them subvocally.
    c. it is more difficult to understand the spoken (or sung) word than the written word.
    d. the ability to write words depends on the strength of the "auditory image" that is evoked.

8. People with phonological dysgraphia _____ words and then write them.

    a. visually imagine
    b. sound out
    c. finger-spell
    d. rehearse

9. People with orthographic dysgraphia have difficulty spelling and writing

    a. compound words.
    b. nonwords.
    c. regular words.
    d. irregular words.

10. Researchers who examined the brains of deceased people with developmental dyslexia found abnormalities in the

    a. corpus callosum.
    b. arcuate fasciculus.
    c. planum temporale.
    d. angular gyrus.

11. Researchers have noted a relationship between developmental dyslexias,

    a. right-handedness, and muscular coordination.
    b. left-handedness, and immune disorders.
    c. right-handedness, and speech difficulties.
    d. ambidexterity, and longevity.

12. All of the following support the hypothesis that dyslexia can result from abnormal input to the parietal lobe caused by an abnormal magnocellular system, except:

a. There is evidence of disorganized magnocellular layers in dyslexics.

b. Dyslexics have trouble with spatial perception and of movements in space.

c. Differences have been recorded in the primary visual cortex of dyslexics.

d. Dyslexics complain that when they try to read, letters seem to move around.

Answers for Self Tests

Lesson I

1. b Obj. 16-1
2. d Obj. 16-1
3. d Obj. 16-2
4. a Obj. 16-2
5. a Obj. 16-3
6. d Obj. 16-3
7. b Obj. 16-3
8. c Obj. 16-4
9. a Obj. 16-5
10. b Obj. 16-5
11. d Obj. 16-5
12. b Obj. 14-5

Lesson II

1. a Obj. 16-6
2. a Obj. 16-6
3. b Obj. 16-7
4. c Obj. 16-7
5. c Obj. 16-7
6. d Obj. 16-7
7. b Obj. 16-8
8. a Obj. 16-8
9. d Obj. 16-8
10. c Obj. 16-9
11. b Obj. 16-9
12. c. Obj. 16-9

CHAPTER 17
Schizophrenia and the Affective Disorders

Lesson I: The Physiology of Schizophrenia

Read the interim summary on page 525 of your text to re-acquaint yourself with the material in this section.

> *Learning Objective 17-1* Describe the symptoms of schizophrenia and discuss the evidence that some forms of schizophrenia are heritable.

Read pages 510-512 and answer the following questions.

1. Schizophrenia affects what percent of the world's population?

2. Who first used the term schizophrenia? What does the term mean? What did he intend the term to signify?

3. Explain the general difference between the positive and negative symptoms of schizophrenia.

4. Describe these positive symptoms.

    a. thought disorders

    b. delusions of persecution, grandeur, and control

    c. hallucinations

5. Describe some of the negative symptoms.

6. Researchers use both twin studies and adoption studies to study the heritability of schizophrenia. What do adoption studies indicate about the incidence of schizophrenia among the children of schizophrenic parents raised by their parents or by nonschizophrenic adoptive parents? (Kety et al., 1968, 1994)

7. What do twin studies indicate about the concordance rate for schizophrenia in monozygotic and dizygotic twins. Use the terms *concordant* and *discordant* in your answer. (Gottesman and Shields, 1982; Tsuang et al., 1991)

8. a. If schizophrenia is a simple trait produced by a single dominant gene, what percentage of the offspring of two schizophrenic parents would be schizophrenic?

    b. And if schizophrenia is a recessive trait, what percentage of the offspring of two schizophrenic parents would be schizophrenic?

    c. What is the actual incidence of this disease among these children?

    d. What does the incidence suggest about the transmission of schizophrenia?

9. a. If a susceptibility to schizophrenia is inherited, which group of people may carry the "schizophrenia gene" but not express it?

    b. Study Figure 17.1 in your text and compare the incidence of schizophrenia in the offspring of monozygotic and dizygotic twins both discordant for the disease. (Gottesman and Bertelsen, 1989)

    c. What do the incidence rates suggest about the heritability of schizophrenia and the inevitability of developing the disease?

10. Describe some of the research on the possible location of the "schizophrenia gene." What can we conclude from these studies? (For example, Dawson and Murray, 1996.)

Learning Objective 17-2 Discuss drugs that alleviate or produce the positive symptoms of schizophrenia; discuss research into the nature of a possible dopamine abnormality in the brains of schizophrenics.

Read pages 512-517 and answer the following questions.

1. State the dopamine hypothesis of schizophrenia in your own words.

2. In general, how do antipsychotic drugs such as chlorpromazine affect schizophrenia?

3. a. What is the pharmacological effect common to all drugs that relieve the positive symptoms of schizophrenia? produce the positive symptoms of schizophrenia?

    b. Names three drugs that produce the positive symptoms of schizophrenia.

4. Most researchers believe which part of the brain is most likely involved in schizophrenia?

5. Why might overactivity of dopaminergic synapses in the nucleus accumbens and amygdala produce the symptoms of schizophrenia? Cite research (Snyder, 1974; Fibiger, 1991) to support your answer.

6. List the three possibilities of abnormalities in dopamine transmission in the brains of schizophrenic patients. (See Table 17.1 in your text.)

   1.

   2.

   3.

7. What does research suggest about the first possibility? Site research by Laruelle et al. (1996) in your answer (See Figure 17.2 in your text.)

8. Describe research to investigate another possibility—that there is an overabundance of dopamine receptors in the brain.

   a. What research methods have been used to study this possibility?

   b. Briefly summarize the inconclusive results.

   c. What were some of the shortcomings of these studies?

   d. What part of the brain did early studies focus on? Why? Why is this the wrong part to study?

9. Explain why the effects of the drug clozapine suggest that the nucleus accumbens is more likely to be involved in schizophrenia that the neostriatum. (Kinon and Lieberman, 1966)

10. Review recent research by Murray et al. (1995) and Gurevich et al. (1997) on D_3 and D_4 receptors in the brains of schizophrenics.

    a. Generally, what did Murray and his colleagues find?

    b. What does the absence of messenger RNA for the D_4 receptors in the nucleus accumbens indicate?

    c. What did the investigators conclude?

d. What did Gurevich et al. (1997) find with regard to D_3 receptors? (See Figure 17.3 in your text.)

11. a. What percentage of schizophrenic patients are not helped by antipsychotic drugs?

    b. What are some common (but usually temporary) side effects of many antipsychotic drugs?

12. a. Describe the disorder tardive dyskinesia and who it affects.

    b. If tardive dyskinesia is produced by an overstimulation of dopamine receptors, why should it be caused by antipsychotic drugs, which are dopamine antagonists?

13. a. Why is clozapine referred to as an "atypical" antipsychotic drug?

    b. What are some advantages and disadvantages of clozapine?

Learning Objective 17-3   Discuss evidence that the negative symptoms of schizophrenia may result from brain damage.

Read pages 517-520 and answer the following questions.

1. Explain why the negative symptoms of schizophrenia suggest that the disease may result from brain damage.

2. List some of the neurological symptoms suffered by schizophrenics. (Stevens, 1982)

3. Let's consider the direct evidence of brain damage.

    a. Using CT and MRI scans, what did Weinberger and Wyatt (1982) discover about the relative ventricle size of schizophrenics and normals of the same age? (Study Figure 17.4 in your text.)

    b. What is the best explanation for the difference in size?

4. a. What other kinds of abnormalities have been documented in later studies?

    b. What parts of the brain appear to be affected? (For example, Bogerts et al., 1993)

    c. What was the relationship, observed in several studies, between severity of brain abnormalities and severity of negative symptoms?

Now let's review epidemiological studies of schizophrenia.

5.   What six environmental factors have been associated with the incidence of schizophrenia?

6.   a.   People whose birthdays fall during the late _____ and early _____ months are more likely to develop schizophrenia later in life. (See Figure 17.5 in your text. Kendell and Adams, 1991). What is this relationship called?

      b.   What threat to a pregnant woman's health rises during these months?

      c.   If the pregnant woman does contract a viral illness, how might her developing fetus be injured?

7.   What is the relationship between population density and the incidence of schizophrenia?

8.   a.   In general, how is the birth rate of babies who later develop schizophrenia affected by influenza epidemics?

      b.   Why does the second trimester of pregnancy appear to be the period of greatest susceptibility? (Sham et al., 1992. See Figure 17.6 in your text.)

9.   Describe the latitude effect and provide a possible explanation.

10.   What did researchers (Susser and Lin, 1992; Susser et al., 1996) find when they studied the offspring of women who were pregnant during the Hunger Winter? Explain the role of a thiamine deficiency in your answer. (Davis and Bracha, 1996)

11.   Finally, what might explain a higher incidence of schizophrenia among the children of women who learned that their husbands had been killed in combat in World War II? (Huttunen and Niskanen, 1978)

Learning Objective 17-4   Describe direct evidence that schizophrenia is associated with brain damage.

Read pages 520-525 and answer the following questions.

1.   Describe the findings of Walker et al. (1994, 1996) using home movies from families with a schizophrenic child.

2.   a.   What happens during the second trimester of pregnancy that makes it a critical time for brain development?

      b.   What did Akbarian et al. (1996) find in the brains of twenty deceased schizophrenics and to what did they attribute the cause?

3. Look at Figure 17.7 in your text and describe the difference in the brains of monozygotic twins that were discordant for schizophrenia. (Suddath et al., 1990)

4. In what case is the prenatal environment of monozygotic twins not identical? (See Figure 17.8 in your text.)

5. Let's review research by Davis, Phelps, and Bracha (1995) on monozygotic twins discordant and concordant for schizophrenia.

   a. What indices did they use to estimate whether twins were monochorionic or dichorionic? Summarize their findings.

   b. Why did Bracha et al. (1992) also examine the fingerprints of monozygotic twins discordant or concordant for schizophrenia? What did they find?

6. Besides the disturbance of normal prenatal development, what other causes can produce schizophrenia? (For example, O'Callaghan et al., 1992)

7. What are some possible explanations for why the symptoms of schizophrenia rarely occur before late adolescence or early adulthood? (Squires, 1977; Torrey, 1991)

8. Weinberger (1988) suggested that the negative symptoms of schizophrenia are caused primarily by damage to the

   _____ _____ _____. The WCST demonstrates that one of the

   functions of this part of the brain is _____ _____. (See Figure 17.9 in your text.)

9. a. What did Weinberger, Berman, and Zec (1986) find with regard to the blood flow of normal subjects performing the WCST test? schizophrenic subjects? (See Figure 17.10 in your text.)

   b. What did Taylor's (1996) review of the literature find?

10. What do Weinberger and his colleagues suggest as an explanation for this "hypofrontality?" What evidence supports this?

11. When Weinberger et al. (1992) took MRI scans and PET scans of the brains of monozygotic twins discordant for schizophrenia, what differences did they observe? How are their observations related to hypofrontality?

12. What have the following researchers suggested to explain the relationship between hypoactivity of the prefrontal cortex and the positive symptoms of schizophrenia, which appear to be produced by the hyperactivity of dopaminergic synapses in the nucleus accumbens?

    a. Weinberger, Berman, and Zec (1986)

    b. Grace (1991) (See Figure 17.11 in your text.)

Lesson I Self Test

1. If the tendency to develop schizophrenia is heritable

   a. the percentage of dizygotic twins concordant for schizophrenia will be higher than that of monozygotic twins.
   b. the incidence of schizophrenia in adopted children with biological schizophrenic parents will be higher than that of the general population.
   c. environment plays no role in the development of the disease.
   d. the disease will be milder if only one biological parent is schizophrenic.

2. The fact that not all children of two schizophrenic parents become schizophrenic suggests that

   a. schizophrenia results from a single faulty gene.
   b. having schizophrenic parents may increase susceptibility to the disease.
   c. the "schizophrenia gene" is recessive.
   d. schizophrenia is not caused by genetic factors.

3. All drugs that relieve the positive symptoms of schizophrenia

   a. are dopamine agonists.
   b. block dopamine reuptake.
   c. are dopamine antagonists.
   d. block dopamine autoreceptors.

4. Which of the following does not support the notion that reinforcement pathways play a role in schizophrenia?

   a. Schizophrenics report feelings of elation and euphoria at the beginning of a schizophrenic episode.
   b. Cocaine and amphetamine release dopamine in reinforcement pathways and can cause psychotic symptoms.
   c. One of the most effective antipsychotic drugs, clozapine, inhibits dopamine in the nucleus accumbens which is part of the reinforcement circuit.
   d. The neostriatum probably plays a more important role in schizophrenia than parts of the brain involved in reinforcement.

5. Tardive dyskinesia

   a. is found in 40 percent of patients receiving antipsychotic medications.
   b. is most common after treatment with the drug clozapine.
   c. is a parkinsonian side effect of antipsychotic medications.
   d. may be due to dopamine receptor supersensitivity.

6. Careful study of CT and MRI scans indicates that schizophrenics have

   a. enlarged ventricles.
   b. fewer convolutions of the cortex.
   c. atrophy of the neostriatum.
   d. a thicker corpus callosum.

7. Tissue or brain abnormalities have been observed in all of the following areas of schizophrenics' brains except:

   a. frontal lobes

b. medial diencephalon
c. cerebellum
d. temporal lobes

8. The "seasonality effect" in schizophrenia refers to

a. the increased likelihood that people born in the summer months develop schizophrenia.
b. the increased likelihood that people born in the winter months develop schizophrenia.
c. the intensification of schizophrenic symptoms in the winter months.
d. the increased incidence of schizophrenia in countries farther from the equator.

9. Which of the following supports the hypothesis that schizophrenia involves abnormal prenatal brain development?

a. NADPH-d distribution in the frontal and temporal lobes was abnormal in a group of deceased schizophrenics.
b. MRI scans show smaller ventricles in the brains of monozygotic twins who are discordant for schizophrenia.
c. The left hippocampus does not develop as rapidly in schizophrenics compared to nonschizophrenics.
d. The fingerprint patterns of twins discordant for schizophrenia are nearly identical.

10. On tests of frontal lobe function, such as the Wisconsin Card Sort Test, schizophrenics *do not:*

a. perform as poorly as people with damage to the dorsolateral prefrontal cortex.
b. show evidence of "hypofrontality."
c. show increased blood flow to the frontal lobes while performing this task.
d. have difficulty changing tasks.

11. Weinberger and colleagues propose that hypofrontality may be caused by

a. increased dopamine input to the frontal lobes.
b. decreased dopamine input to the frontal lobes.
c. increased dopamine input to the temporal lobes.
d. decreased dopamine input to the temporal lobes.

12. Hyperactivity of dopaminergic neurons in the nucleus accumbens

a. accounts for the negative symptoms of schizophrenia.
b. results from high activity of glutamate neurons that originate in the prefrontal cortex.
c. involves increased receptor sensitivity in response to decreased tonic released of dopamine.
d. is enhanced by drugs like amperozide.

Lesson II: The Major Affective Disorders

Read the interim summary on page 535 of your text to re-acquaint yourself with the material in this section.

Learning Objective 17-5 Describe the two major affective disorders, the heritability of these diseases, and their physiological treatments.

Read pages 526-529 and answer the following questions.

1. The primary symptom of schizophrenia is _____ _____ and the primary symptom of the major affective disorders is _____ _____.

2. Describe bipolar disorder.

a. Name and describe the two alternating moods.

b. Approximately how long is each episode?

c. What is the incidence of bipolar disorder in men and women?

3. Describe unipolar depression, including its incidence in men and women.

4. Present a profile of patients suffering from affective disorders by describing the effect of the disease on
    a. self-esteem.

    b. personal safety.

    c. energy level.

    d. appetite.

    e. sex drive.

    f. sleep patterns.

    g. body functions.

5. Explain the differences between mania and a normal enthusiasm for life.

6. State the results of these studies tracing the heritability of the affective disorders.
    a. incidence among close relatives of patients (Rosenthal, 1971)

    b. incidence in sets of monozygotic and dizygotic twins (Gershon et al., 1976) reared together or apart (Price, 1968)

    c. the existence of a gene responsible for susceptibility to bipolar disorder. (Spence et al., 1995; Kelsoe et al., 1996)

7. List four effective biological treatments of depression.

    1.   3.

    2.   4.

8. How is bipolar disorder treated?

9. a. Explain how drugs such as iproniazid affect the brain. Be sure to state the function of monoamine oxidase (MAO).

    b. Describe the most common serious side effect of MAO inhibitors.

10. Explain the pharmacological effects of the tricyclic antidepressant drugs.

11. Briefly describe the history of ECT.

12. a. Compare the speed with which MAO inhibitors and ECT relieve depression.

    b. State a serious side effect of the excessive use of ECT. (Squire, 1974)

    c. State two reasons why occasional use of ECT may be justified. (Baldessarini, 1977)

13. a. Which phase of bipolar disorder is effectively treated with lithium? (Gerbino et al., 1978)

    b. What are some of the significant advantages of treating bipolar disorder with lithium? (Fieve, 1979)

    c. State a tentative explanation of the effect of lithium on the brain. (Atack et al., 1995; Jope et al., 1996)

    d. What is an alternative medication to lithium? How effective is it? (See Figure 17.13 in your text. Post et al., 1992)

Learning Objective 17-6 Summarize the monoamine hypothesis of depression and review the long-term changes in receptor sensitivity.

Read pages 529–532 and answer the following questions.

1. Briefly explain the monoamine hypothesis of depression.

2. Begin a review of evidence that supports this hypothesis. When reserpine was first used to treat high blood pressure, what side effect did physicians notice in about 15 percent of their patients? (Sachar and Baron, 1979)

3. Explain how reserpine affects the membrane of synaptic vesicles and the release of monoamine transmitter substance in the brain.

4. Explain how the pharmacological and behavioral effects of reserpine complement those of MAO inhibitors.

5. Describe research on CSF levels of 5-HIAA and suicidal depression.

   a Explain how 5-HIAA is produced in the brain.

   b. Compare the CSF levels of 5-HIAA in the brains of people who

      1. had attempted suicide and control subjects.

      2. were depressed and eventually committed suicide and those who did not.(Träskmann et al, 1981; Roy et al., 1989)

   c. What did analysis of the CSF of healthy, nondepressed volunteers reveal? (Sedvall et al., 1980)

6. a. How did Delgado et al. (1990) lower the tryptophan level in subjects' brains?

   b. What effect did the treatment have on the level of serotonin in the brain? Why?

   c. What was the effect of the experiment on subjects' depression?

7. What are the results of the administration of AMPT? (See Figure 17.14 in your text. Heninger et al., 1996)

8. a. Summarize an explanation of why antidepressant drugs do not relieve symptoms for many days. (Sulser and Sanders-Bush, 1989.)

   b. Now summarize another explanation, proposed by Artigas et al. (1996).

9. Carefully study Figure 17.15 in your text and describe the hypothesized process illustrated there.

10. If this explanation is correct, what should be the effect of drugs that block these receptors? Cite research to support your answer. (Artigas et al, 1993; Blier and Bergeron, 1995.)

11. What is the effect of antidepressant drugs on the sensitivity of D_1 and D_2 dopamine receptors in the nucleus accumbens?

12. a. What evidence is there of brain abnormalities in depressed patients? (Soares and Mann, 1997; Elkis et al., 1996; Drevets et al., 1992; Wu et al., 1992).

   b. Study Figure 17.16 in your text and describe the metabolic changes before and after ECT treatment. (Nobler et al., 1994)

Learning Objective 17-7 Explain the role of circadian and seasonal rhythms in affective disorders: the effects of REM sleep deprivation and total sleep deprivation, and seasonal affective disorder.

Read pages 532-535 and answer the following questions.

1. Describe how the sleep of people with depression is disrupted.

   a. amounts of slow-wave delta sleep and stage 1 sleep

   b. fragmentation of sleep

   c. changes in REM sleep patterns (See Figure 17.17 in your text.)

2. a. What is the effect of selective deprivation of REM sleep? (Vogel et al., 1975, Vogel et al., 1990)

   b. Approximately how long does it take for relief from depression to occur?

3. a. How did antidepressant drugs effect the sleep cycle of cats? (Scherschlicht et al., 1982)

   b What is the physiological effect of all drugs that suppress REM sleep? (Vogel et al., 1990)

c. However, why is it incorrect to conclude that all antidepressant drugs relieve depression by suppressing REM sleep?

4. a. What change in REM sleep was observed in first-degree relatives of people with depression? (Giles et al., 1987)

    b. Which family members had the highest risk of becoming depressed? (Giles et al., 1988)

    c. What did a comparison of REM sleep patterns of the newborn infants of mothers with and without a history of depression indicate? (Coble et al., 1988)

5. Briefly describe some of the characteristics of the animal model of depression, developed by Vogel et al. (1990).

6. Describe research on the antidepressant effect of total sleep deprivation.

    a. How quickly does total sleep deprivation relieve depression? (See Figure 17.18 in your text. Wu and Bunney, 1990)

    b. According to Wu and Bunney (1990), why does sleep trigger depression in susceptible people?

    c. Now study their supporting evidence shown in Figure 17.19 in your text, which is a summary of the results of eight different studies, and describe how the self-ratings of depression changed over a three day period.

7. a. What characteristic of a depressed patient predicts a good response to total sleep deprivation? (Reinink et al., 1990; Haug, 1992)

    b. Look again at Figure 17.19 and note how the mood of people who responded to sleep deprivation treatment changed during the day.

8. a. How quickly and how long does REM sleep deprivation relieve depression? total sleep deprivation?

    b. What effect can total sleep deprivation have on people with bipolar disorder? (Wehr, 1992)

c. These findings suggest that what kind of research might be profitable?

9. a. What is a therapeutic benefit of partial sleep deprivation or intermittent sleep deprivation? (Szuba et al., 1991; Leibenluft and Wehr, 1992)

b. What is the best method of partial sleep deprivation to follow?

10. Describe the symptoms of seasonal affective disorder, noting how they differ from the symptoms of major depression.

11. a. Outline a possible explanation why phototherapy is effective against seasonal affective disorder. Be sure to use the term *zeitgeber* in your answer. (Rosenthal et al., 1985; Stinson and Thompson, 1990)

b. Does it matter what time of day light therapy occurs? What does this fact suggest? (Wirz-Justice et al., 1993; Meesters et al., 1995)

12. What percentage of people are sensitive to seasonal changes in the hours of sunlight? What is an easy treatment for "winter blahs?"

Lesson II Self Test

1. Bipolar disorder is characterized by

   a. unremitting or episodic depression without periods of mania and afflicts more women than men.
   b. mania without periods of depression and afflicts more women than men.
   c. alternating bouts of depression followed by periods of normal affect and afflicts more men than women.
   d. alternating bouts of mania and depression and afflicts men and women about equally.

2. Studies of the genetic basis of affective disorders have shown that

   a. there is little evidence in favor of a genetic component in these disorders.
   b. concordance rates for monozygotic twins are considerable higher than concordance rates for dizygotic twins.
   c. there is a genetic basis for unipolar depression but not bipolar disorder.
   d. a gene on chromosome 11 is definitely involved in bipolar disorder.

3. MAO inhibitors are not routinely used to treat depression because they

   a. cause memory loss.
   b. increase pressor amines (the cheese effect).
   c. are simply not as effective as other drugs.
   d. cause hypotension (lowered blood pressure).

4. A serious side effect of frequent electroconvulsive therapy is

   a. disturbances in biological rhythms.
   b. elevated blood pressure.
   c. long-lasting memory impairments.
   d. suppression of normal feelings of emotion.

5. All of the following are true of lithium except:

   a. It is typically administered during the manic phase of bipolar disorder.
   b. It does not alter normal emotions or intellectual functioning.
   c. It may act in the brain by stabilizing populations of receptors for certain transmitter substances.
   d. It is a very effective treatment, so compliance is not a problem.

6. The monoamine hypothesis suggests that depression is a result of _____ of monoaminergic neurons.

   a. insufficient activity
   b. excessive numbers
   c. overactivity
   d. the proliferation

7. Depressed individuals fed a diet low in tryptophan and a cocktail high in other amino acids

   a. become manic.
   b. relapse into depression.
   c. show changes in cognition but not affect.
   d. have elevated levels of serotonin metabolites.

8. Subsensitivity of postsynaptic noradrenergic receptors

   a. occurs two to three weeks after symptoms of depression improve.
   b. follows treatment of depression with reserpine.
   c. occurs immediately after starting antidepressant treatment.
   d. occurs following treatment with antidepressant drugs, ECT, and sleep deprivation.

9. Which two neurotransmitters may play the greatest role in depression?

   a. dopamine and norepinephrine
   b. dopamine and serotonin
   c. norepinephrine and serotonin
   d. dopamine and glutamate

10. People whose depression is relieved by total sleep deprivation

   a. begin to feel better immediately.
   b. feel better in the morning than in the evening.
   c. must limit their sleep to brief naps.
   d. gradually feel better over the course of several weeks.

11. Wu and Bunney suggested that sleep deprivation causes an improvement in depressive symptoms because

   a. during sleep a depressogenic substance is produced that needs to be metabolized during waking hours.
   b. waking produces a substance with antidepressant effects.
   c. REM sleep allows a person to actively rehearse the life events that may be causing depression.
   d. it causes memory loss.

12. An effective treatment for seasonal affective disorder is

   a. avoidance of temperature changes, especially at night.
   b. infrequent naps to regulate the amount of REM sleep.
   c. a fixed meal schedule to minimize changes in metabolic rate.
   d. exposure to several hours of bright light each day.

Answers for Self Tests

Lesson I

1. b Obj. 17-1
2. b Obj. 17-1
3. c Obj. 17-2
4. d Obj. 17-2
5. d Obj. 17-3
6. a Obj. 17-3
7. c Obj. 17-3
8. b Obj. 17-3
9. a Obj. 17-4
10. c Obj. 17-4
11. b Obj. 17-4
12. c Obj. 17-4

Lesson II

1. d Obj. 17-5
2. b Obj. 17-5
3. b Obj. 17-5
4. c Obj. 17-5
5. d Obj. 17-5
6. a Obj. 17-6
7. b Obj. 17-6
8. d Obj. 17-6
9. c Obj. 17-6
10. a Obj. 17-7
11. a Obj. 17-7
12. d Obj. 17-7

CHAPTER 18
Anxiety Disorders, Autistic Disorder, and Stress Disorders

Lesson I: Anxiety Disorders and Autistic Disorder

Read the interim summary on page 544 of your text to re-acquaint yourself with the material in this section.

Learning Objective 18-1 Describe the symptoms and possible causes of panic disorder.

Read pages 538-540 and answer the following questions.

1.   What is the primary symptom of the anxiety disorders?

2.   a. What is the incidence of panic disorder in the general population? (Robbins et al., 1984)

   b. At what age does this disorder most commonly begin? (Woodruff et al., 1972)

3.   a. Describe a panic attack.

   b. Describe the relation between panic attacks and agoraphobia. Use the term *anticipatory anxiety* in your answer.

4.   Why are people who suffer from panic disorder often convinced it is a medical, rather than a mental, condition?

5.   a. What is the incidence of panic disorder for monozygotic and dizygotic twins? (Slater and Shields, 1969) for first-degree relatives of people with panic disorder? (Crowe et al., 1983)

   b. What does the familial pattern of panic disorder suggest about its origin? (Crowe et al., 1987)

6.   How may a panic attack be induced? (Stein and Uhde, 1994)

7. a. Describe how Roth et al. (1992) tested the suggestion that people with panic attacks have more reactive autonomic nervous systems.

    b. What results did their experiment yield?

8. a. After receiving an injection of sodium lactate, what percentage of a group of normal subjects experienced a panic attack? (Balon et al., 1989)

    b. What percentage of the relatives of the subjects who experienced a panic attack had a history of anxiety disorders?

9. What are the two components of the treatment for anxiety disorders?

    1.   2.

10. Explain how benzodiazepines affect the brain.

11. What is the effect of drugs that *reduce* the sensitivity of the GABA binding site?

12. a. What possible causes of anxiety disorders are suggested by the effects of drugs on GABA receptors?

    b. What other substances may be involved? Cite research to support your answer. (Bradwejn et al., 1990; Adams et al., 1995; Coplan et al., 1992)

Learning Objective 18-2 Describe the symptoms and possible causes of obsessive-compulsive disorder.

Read pages 540-544 and answer the following questions.

1. Define *obsessions* and *compulsions* in your own words.

2. a. What is the incidence of obsessive-compulsive disorder

        1. in the general population?

        2. among men and women?

    b. At what age does this disorder commonly begin? (Robbins et al., 1984)

c. What are the characteristics of this disorder in various racial and ethnic groups? (Akhtar et al., 1975; Khanna and Channabasavanna, 1987; Hinjo et al., 1989)

3. List the four categories of compulsions and give an example of each. (Study Table 18.1 in your text.)

    1.

    2.

    3.

    4.

4. What is the possible relationship between compulsive behaviors and species-typical behaviors? (Wise and Rapoport, 1988)

5. Fiske and Haslam (1997) propose another explanation for the origins of obsessive-compulsive disorder. Summarize their hypothesis.

6. Review research on the possible causes of hereditary obsessive-compulsive disorder.

    a. What have family studies revealed about the incidence of obsessive-compulsive disorder and Tourette's syndrome? (Pauls and Leckman, 1986; Pauls et al., 1986)

    b. Briefly describe Tourette's syndrome especially any similarities with obsessive-compulsive disorder. (Leonard et al., 1992a, 1992b)

    c. What may be the cause of both obsessive-compulsive disorder and Tourette's syndrome?

7. Now review research on the possible causes of nonhereditary obsessive-compulsive disorder.

    a. What injuries or diseases may contribute to this disorder? (Berthier et al., 1966; Hollander et al., 1990)

    b. Which brain regions appear to be damaged or dysfunctional? (Giedd et al., 1995; Robinson et al., 1995)

    c. Describe Sydenham's chorea and its relationship to obsessive-compulsive disorder. (Swedo et al., 1989a; Cummings and Cunningham, 1992; Husby et al., 1976)

8. a. What have PET scans indicated about the brains of patients with obsessive-compulsive disorder? (Baxter et al., 1989; Swedo et al., 1989b; Rubin et al., 1992; Lucey et al., 1997)

    b. Describe research strongly implicating the prefrontal cortex in obsessive-compulsive disorder. (Swedo et al., 1992; Rubin et al., 1995, Schwartz et al., 1996.)

9. a. What interesting technique did Breiter et al. (1996) use to investigate the brain functions of people with obsessive-compulsive disorder?

    b. What did they find?

10. a. What kind of surgery is often a successful treatment of this disorder? (Ballantine et al., 1987; Mindus et al., 1994)

    b. Under what circumstances is surgery justified?

11. The most effective treatment of obsessive-compulsive disorder is _____ _____.

12. Study Figure 18.1 in your text and compare the effectiveness of clomipramine and an antidepressant drug desipramine in relieving the symptoms of obsessive-compulsive disorder. Explain what happened after the switch. (Leonard et al., 1989)

13. a. How do all effective antiobsessional drugs affect the brain? What does this fact suggest about the cause of obsessive-compulsive disorder?

    b. Briefly describe several compulsive behaviors which afflict people and how they are successfully treated. (Leonard et al., 1992)

    c. Now describe a compulsive behavior seen in some breeds of large dogs and how it can be successfully treated. (Rapoport et al., 1992)

Read the interim summary on page 548 of your text to re-acquaint yourself with the material in this section.

Learning Objective 18-3   Describe the symptoms and possible causes of autism.

Read pages 544-548 and answer the following questions.

1. What is the incidence of autism in the general population? between boys and girls?

2. List three key symptoms of autistic disorder. Illustrate each symptom with specific examples.

    1.

    2.

    3.

3. a. According to Frith and her colleagues, what is the major deficit suffered by individuals with autistic disorder?

    b. Describe the experiment by Baron-Cohen et al. (1985) that supports this hypothesis. (See Figure 18.2 in your text.)

4. a. Describe the historical (and incorrect) approach to understanding the basis of autism.

    b. What is the modern approach to investigating the possible causes of autism?

5. a. Cite evidence from family studies, especially the study of autism in siblings, that supports the role of genetics in this disorder. (Folstein and Piven, 1991; Bailey, 1993)

    b. Compare the concordance rates for autism in monozygotic and dizygotic twins. (Folstein and Piven, 1991; Bailey et al., 1995)

    c. What is unique about the twin who develops autism in monozygotic pairs who are discordant for autism? What does this suggest about the causes of this disorder?

6. Phenylketonuria (PKU) has been linked to autism. Describe this disorder, how it affects brain development, and how it may be treated.

7. a. Why do researchers believe there may be a genetic link between autism and Tourette's syndrome? (Comings and Comings, 1991)

b. What did Sverd et al. (1991) find in the family histories of patients with symptoms of autism and Tourette's syndrome? (See Table 18.2 in your text.)

8. What is the probable link between autism and fragile X syndrome? (Reiss and Freund, 1992; Fisch, 1992)

9. Cite several incidents in prenatal development that may lead to autism. (Chess et al., 1971; Fernell et al., 1991; Strömland et al., 1994; Ritvo et al., 1990).

10. Where have researchers found evidence of abnormalities in the brains of autistic people? (See Figure 18.3 in your text. DeLong, 1992; Happé and Frith, 1996; Courchesne, 1991; Holroyd et al., 1991; Hashimoto et al, 1995).

Lesson I Self Test

1. People with panic disorder suffer from anticipatory anxiety, which is

   a. a brief interval of unrealistic fear that precedes a panic attack.
   b. the first stage of a panic attack.
   c. often sufficient to trigger a panic attack.
   d. the fear another panic attack will strike.

2. Studies concerning the genetic basis of panic disorder have shown

   a. that 50 percent of the first degree relatives of a person with panic disorder also have panic disorder.
   b. the concordance rate for monozygotic twins is similar to that found in dizygotic twins.
   c. that panic disorder may be caused by a single, dominant gene.
   d. that environmental factors outweigh genetic factors in the etiology of this disorder.

3. An injection of _____ may cause a panic attack in susceptible people.

   a. alcohol
   b. lactic acid
   c. benzodiazepine
   d. lithium carbonate

4. All of the following substances have been implicated in panic disorder except:

   a. epinephrine
   b. serotonin
   c. cholecystokinin
   d. benzodiazepine

5. People suffering from obsessive-compulsive disorder

   a. discuss their behavior openly.
   b. have periodic episodes of mania.
   c. recognize that their thoughts and behaviors are senseless.
   d. are often insomniacs.

6. Obsessive-compulsive disorder has been linked to all of the following except:

   a. exposure to mumps
   b. Tourette's syndrome
   c. birth trauma
   d. encephalitis

7. PET scans have recorded abnormal activity in all of the following regions of the brains of patients with obsessive-compulsive disorder except:

   a. prefrontal cortex
   b. cingulate cortex
   c. basal ganglia
   d. cerebellum

8. All effective antiobsessional drugs

a. block dopamine receptors.
b. are MAO antagonists.
c. increases the sensitivity of the GABA binding site.
d. block the reuptake of 5-HT.

9. Which is not true of the language of autistic children?

   a. It is abnormal or even nonexistent.
   b. It often includes repetition of what others have said.
   c. It improves in late adolescence.
   d. It is often self-centered or self-interested.

10. Frith's model of autism as the inability to see the world from others' points of view does not predict which of the following symptoms of the disorder?

    a. abnormal social relationships
    b. stereotyped movements

c. impaired imagination
d. inability to predict and explain other people's behavior

11. The best evidence for the genetic basis of autism comes from

    a. twin studies showing near-100 percent concordance for monozygotic twins.
    b. examination of the extended family tree.
    c. family studies which show that most affected families have more than one autistic child.
    d. all of the above.

12. Nongenetic causes of autism include

    a. hydrocephalus.
    b. PKU.
    c. exposure to cytomegalovirus.
    d. elevated levels of CCK.

Lesson II: The Stress Disorders

Read the interim summary on pages 559 in your text to re-acquaint yourself with the material in this section.

Learning Objective 18-4 Describe the physiological responses to stress and their effects on health.

Read pages 549-552 and answer the following questions.

1. Define *stress* in your own words.

2. Explain why long-term stress has harmful effects on health. Be sure to use the phrase *fight-or-flight response* in your answer.

3. Which components of emotion are responsible for the harmful effects of health? Why?

4. _____ and _____ responses are catabolic and ready the body's _____

   _____.

5. a. Where is epinephrine secreted?

   b. What are some of its effects? Which effect of epinephrine and norepinephrine together contributes to cardiovascular disease?

c. In stressful situations, what change occurs in the secretion of norepinephrine in the brain? (Yokoo et al., 1990; Cenci et al., 1992)

d. If noradrenergic axons from the brain stem to the forebrain are destroyed, how is the response to social isolation stress affected? (Montero et al., 1990)

e. Trace the brain circuit that appears to produce the release of norepinephrine. (Wallace et al., 1992)

6. a. What is the other stress-related hormone and where is it secreted?

b. List some of its effects.
   1.   4.
   2.   5.
   3.   6.

c. What is the presumed significance of the fact that nearly every cell of the body has glucocorticoid receptors?

d. The secretion of glucocorticoids is controlled by neurons of the _____ _____ of the _____ which secrete a peptide called _____-_____ _____, which, in turn, stimulates the anterior pituitary gland to secrete _____ _____. The _____ _____ secretes glucocorticoids in response to ACTH.

The control of the secretion of glucocorticoids is illustrated in Figure 18.4 in your text.

7. a. Where does CRF serve as a neuromodulator/neurotransmitter?

b. What are the effects of an intracerebroventricular injection of CRF? (Britton et al., 1982; Cole and Koob, 1988; Swerdlow et al., 1986) of a CRF antagonist? (Kalin et al., 1988; Heinrichs et al, 1994; Skutella et al., 1994)

8. a. What are the effects of stress on rats who have had their adrenal glands removed?

b. What medical treatment do adrenalectomized human patients receive in times of stress? (Tyrell and Baxter, 1981)

9. Study Figure 18.5 in your text and describe the incidence of hypertension in air traffic controllers in high-stress and low-stress airports.

10. a. According to Selye (1976), what event causes the harmful effects of stress?

    b. List some of the effects on health of prolonged stress.

    c. How were the effects of stress on healing demonstrated? What were the results? (Kiecold-Glaser et al., 1995) (See Figure 18.6 in your text.)

11. a. How does the hippocampus change as a person ages? What is the result?

    b. Describe how long-term exposure to glucocorticoids affects hippocampal neurons and the possible consequences as a person ages. (Sapolsky, 1986; Sapolsky et al., 1986)

    c. What supporting evidence did Lupien et al. (1996) find?

12. a. Briefly describe the social structure of vervet monkey colonies.

    b. What happened to the monkeys who were subjected to intense stress? (See Figure 18.7 in your text. Uno et al., 1989)

    c. Cite evidence that stress induced brain degeneration occurs in humans as well. (Jensen et al., 1982)

13. a. State two effects of maternal prenatal stress on offspring. (Takahashi et al., 1992)

    b. Describe evidence that suggests that the effects of prenatal stress are mediated by the secretion of glucocorticoids. (Barbazanges et al., 1996).

Learning Objective 18-5 Discuss some of the long term effects of stress: posttraumatic stress disorder, cardiovascular disease, and the coping response.

Read pages 552-555 and answer the following questions.

1. a. Describe *posttraumatic stress disorder* in your own words.

    b. Describe some of the symptoms.

    c. At what age does this disorder usually occur?

2. a. What four factors increase the likelihood that a soldier subjected to combat stress would develop posttraumatic stress disorder? (Kulka et al., 1990)

    b. Does any evidence suggest that genetic factors play a role in determining susceptibility to this disorder? (True et al., 1993)

    c. What additional factor associated with susceptibility to posttraumatic stress disorder did Comings et al. (1996) identify?

3. What do MRI studies reveal about the effects of posttraumatic stress disorder on the hippocampus? (Bremner et al., 1995; Gurvits et al., 1996)

4. What did a PET study by Shin et al. (1997) reveal?

5. a. What event occurs when the blood vessels to the heart become blocked? when the blood vessels to the brain become blocked?

    b. List the two most important risk factors for cardiovascular disease

        1.   2.

6. Review research to assess the relationship between cardiovascular disease and individual differences in stress reactions.

    a. Describe the cold pressor test.

    b. What were the results of a comparison of the subjects' reaction to this test as children and the incidence of high blood pressure later in life? (Wood et al., 1984)

    c. Which monkeys who had been subjected to a stressful situation later showed the highest rates of coronary artery disease? (Manuk et al., 1983, 1986)

d. How was the blood pressure of normal rats affected by hypothalamic tissue transplants from hypertensive rats and from normal rats? (See Figure 18.9 in your text. Eilam et al., 1991)

7. Identify the individual differences that may determine the severity of a stressful situation.

8. What is one of the most important variables that determines whether an aversive stimulus will cause a stress reaction?

9. Compare the responses of rats and humans who are both permitted some degree of control in stressful situations. (Weiss, 1968; Gatchel et al., 1989)

10. a. What substance may play a role in coping responses? (Drugan et al., 1994)

    b. Study Figure 18.10 in your text and compare the levels of endogenous benzodiazepines in the brains of the "coping" group and home cage group of rats who were exposed to stress.

Learning Objective 18-6 Discuss psychoneuroimmunology and the interactions between the immune system and stress.

Read pages 555-559 and answer the following questions.

1. Define *psychoneuroimmunology* in your own words.

2. a. Let's construct an overview of the functions of the immune system. The white blood cells of the immune system develop in the _____ _____ and the _____ _____.
   Some of these cells circulate in the _____ or _____ _____ and others reside permanently in one _____.

   b. What triggers an immune reaction? What are the two types of reactions?

   c. Describe a nonspecific inflammatory reaction resulting from tissue damage.

   d. Describe a nonspecific reaction resulting from tissue infection from a virus.

   e. Explain the role and importance of natural killer cells.

f. Briefly describe a specific immune reaction-the chemically mediated reaction by explaining the relationship between

1. antigens and antibodies.

2. B-lymphocytes and immunoglobulins.

3. immunoglobulins and antigens. (Study Figure 18.11a in your text.)

g. Briefly describe a second specific immune reaction—the cell-mediated reaction. Be sure to explain where antibodies produced by T-lymphocytes are found. (See Figure 18.11b.)

h. What is the role of the cytokines in cell communication and cell division?

i. Where are cytokines produced and what does their release cause?

j. Finally, how may glucocorticoids suppress this immune response? (Sapolsky, 1992)

3. State the general conclusion of research on stress and the immune system of the caregivers of family members with Alzheimer's disease (Kiecolt-Glaser et al., 1987), of husbands who lost their wives to breast cancer (See Figure 18.12 in your text; Schleifer et al., 1983), and of healthy subjects imagining past unpleasant situations (Knapp et al., 1992).

4. a. What is the most important mechanism by which stress impairs the immune system?

b. Keller et al. (1983) found decreased lymphocytes in rats who had experienced the stress of inescapable shock. How did they determine the cause? (See Figure 18.13 in your text.)

5. a. How does the brain control the suppression of the immune system?

b. Explain how the same mechanism that produces negative emotional responses is also responsible for immunosuppression. Cite research to support your answer. (Sharp et al., 1991; Imaki et al., 1992)

6. What other parts of the body may play a role in immunosuppression?

7. a. Describe two effects of inescapable intermittent shock.

    b. Which brain chemicals appear to mediate these effects and how can they be abolished in the laboratory? (Shavit et al., 1984)

    c. How can natural killer cell activity also be suppressed? (Shavit et al, 1986)

8. What stress related responses sometimes occur to

    a. a surviving spouse?

    b. medical students during final examinations? (Glaser et al., 1987)

    c. patients with rheumatoid arthritis? (Feigenbaum et al., 1979)

    d. rats who are handled or exposed to a cat? (Rogers et al., 1980)

    e. rats predisposed to diabetes and subjected to moderate chronic stress? (Lehman et al., 1991)

9. a. Explain the hypothesis of research on stress and upper respiratory illness. (Stone et al., 1987)

    b. Summarize the results of records kept by volunteers who developed upper respiratory illness which are shown in Figure 18.14 in your text.

    c. How did the researchers account for the effect? Be sure to refer to IgA in your answer.

    d. Briefly describe research that confirmed this study by Cohen, Tyrrell, and Smith (1991). (See Figure 18.15 in your text.)

Lesson II Self Test

1. Select the correct statement about stress.

    a. Short-term exposure to stressors typically causes conditions such as ulcers.
    b. The fight-or-flight response is a maladaptive reaction to stressors.
    c. Emotional responses are generally useful and adaptive, but can be hazardous if continuous rather than episodic.
    d. The deleterious effects of stress on health generally have been over-emphasized by the media; recent research does not support this link.

2. Which of the following is not a stress-related hormone released by the adrenal glands?

    a. epinephrine
    b. cortisol
    c. ACTH
    d. norepinephrine

3. The secretion of glucocorticoids is controlled by neurons in the _____ and glucocorticoid receptors are _____.

    a. paraventricular nucleus of the hypothalamus; contained in almost every cell of the body.
    b. central nucleus of the amygdala; found in highest concentration in the adrenal glands.
    c. hippocampal formation; especially susceptible to the effects of stress.
    d. hypothalamus; the first to signal a rise in blood pressure.

4. Long-term stress increases the secretion of _____ which may be responsible for _____.

    a. aldosterone; cardiovascular disease
    b. epinephrine; sodium metabolism
    c. glucocorticoids; the harmful effects of stress
    d. antigens; autoimmune diseases

5. Young vervet monkeys near the bottom of the social hierarchy who experienced almost constant stress

    a. failed to learn normal group coping responses.
    b. engaged in more fight-or-flight responses than other young monkeys who were not subjected to stress.
    c. later showed the highest rates of coronary artery disease.

    d. sustained severe damage to the hippocampal formation.

6. Posttraumatic stress disorder

    a. only strikes in late adolescence or early adulthood.
    b. is sometimes associated with damage to the hippocampal formation.
    c. is sometimes associated with damage to the anterior cingulate gyrus.
    d. rarely occurs in subjects with the A1 allele.

7. Which of the following is not a risk factor for cardiovascular disease?

    a. high blood pressure and high cholesterol
    b. exposure to the cold pressor test
    c. genetic differences in brain chemistry
    d. high emotional reactivity

8. Why did the emotional response of a group of rats to inescapable shock disappear after they learned a coping response?

    a. The pain was diminished.
    b. The number of shocks was reduced.
    c. Their stomach ulcers healed.
    d. They were permitted some control over the situation.

9. The neural mechanisms responsible for coping responses may involve the secretion of

    a. endogenous opioids.
    b. glucocorticoids.
    c. endogenous benzodiazepines.
    d. GABA.

10. The immune system develops _____ through exposure to _____.

    a. antibodies; antigens
    b. interferon; antibodies
    c. antigens; antibodies
    d. antibodies; B-lymphocytes

11. Cytokines

    a. are the body's first defense against malignant tumors.
    b. stimulate cell division.
    c. develop in the bone marrow.
    d. are unique proteins on the surface of infectious microorganisms

12. Which of the following is true of the immunoglobulin, IgA?

    a. High levels of IgA are associated with an unhappy mood in the subject.

    b. Stress has no effect on the production of IgA.

    c. IgA is secreted in the nose, mouth, throat, and lungs, and acts as a defense against infection.

    d. Stress stimulates the production of IgA.

Answers for Self Tests

Lesson I			Lesson II		
1.	d	Obj. 18-1	1.	c	Obj. 18-4
2.	c	Obj. 18-1	2.	c	Obj. 18-4
3.	b	Obj. 18-1	3.	a	Obj. 18-4
4.	a	Obj. 18-1	4.	c	Obj. 18-4
5.	c	Obj. 18-2	5.	d	Obj. 18-4
6.	a	Obj. 18-2	6.	b	Obj. 18-5
7.	d	Obj. 18-2	7.	b	Obj. 18-5
8.	d	Obj. 18-2	8.	d	Obj. 18-5
9.	c	Obj. 18-3	9.	c	Obj. 18-5
10.	b	Obj. 18-3	10.	a	Obj. 18-6
11.	a	Obj. 18-3	11.	b	Obj. 18-6
12.	a	Obj. 18-3	12.	c	Obj. 18-6

CHAPTER 19
Drug Abuse

Lesson I: Definitions and Characteristics of Addiction

Read the interim summary on page 565 in your text to re-acquaint yourself with the material in this section.

Learning Objective 19-1   Review the general characteristics and definitions of addiction.

Read pages 562-565 and answer the following questions.

1. Describe the adverse effects of the following addictive substances:

    a. alcohol

    b. smoking

    c. cocaine

    d. designer drugs

2. Briefly review the history of addictive drugs.

3. How do Eddy et al. (1965) define the following terms?

    a. physical dependence

    b. psychic dependence

4. Define the following terms.

    a. tolerance

    b. withdrawal symptom

5. Carefully explain how the body's compensatory mechanisms to re-establish homeostasis may account for both drug tolerance and the accompanying withdrawal symptoms.

6. a. Explain why withdrawal symptoms are not the reason why someone becomes a drug addict or remains addicted.

   b What is the reason people take drugs?

   c. Why did people neglect the addictive properties of cocaine? Which is more addictive: cocaine or heroin?

   d. Is physiological or psychological addiction more important? Provide reasons for your answer.

7. a. Summarize the World Health Organization's definition of addiction.

   b. What does the *DSM-IV* definition stress?

   c. What does your text conclude is the essential characteristic of an addictive drug?

Read the interim summary on page 571 in your text to re-acquaint yourself with the material in this section.

| *Learning Objective 19-2* Describe two common features of addiction: positive and negative reinforcement. |

Read pages 565-568 and answer the following questions.

1. Drugs that lead to dependency must first reinforce _____ _____.

2. What does the occurrence of an appetitive stimulus activate?

3. a. Describe an experiment that illustrates the importance of the immediacy of reinforcement. (Logan, 1965)

b. What do the results of this experiment tell us about the addictive qualities of different drugs?

c. Why would someone choose to become addicted to a drug that has powerful aversive effects in the long term?

d. Why are nonhuman animals unlikely to become addicted to drugs administered in the form of a pill?

4. What physiological effect is common to all natural reinforcers? (White, 1996)

5. Describe two ways in which drugs can trigger the release of dopamine.

6. What other substance may be involved in reinforcement?

7. Review research by Lamb et al. (1991) that indicates that positive reinforcement can take place without any feelings of pleasure.

a. Who participated in the study?

b. What were the subjects asked to do? What did they receive?

c. How did they describe the effects of the highest does? the lower doses?

d. Compare their lever pressing behavior with their subjective reports. (Study Figure 19.1 in your text.)

8. a. Define *negative reinforcement* in your own words. How does it differ from punishment?

b. Give an example of how a drug can provide negative reinforcement.

Learning Objective 19-3 Describe further common features of addiction: tolerance and withdrawal; craving and relapse.

Read pages 568-571 and answer the following questions.

1. Describe the two types of compensatory mechanisms involved in tolerance. (See Figures 19.2 and 19.3 in your text.)

    1.

    2.

2. a. How did Siegel et al. (1982) demonstrate the potency of classical conditioning as a mechanism for tolerance?

    b. What are the implications of this study for humans?

3. Review explanations for craving from

    a. Robinson and Berridge (1993).

    b. Hyman (1996b).

4. When people with a history of cocaine abuse see a video of drug-related stimuli, what do PET scans reveal? (Grant et al., 1996)

5. Long-term cocaine abuse may increase the sensitivity of _____ _____ _____ and

    their stimulation with _____ is reinforcing.

6. Look at Figure 19.4 in your text and describe what Staley and Mash (1996) found in the brains of people who had died of cocaine overdoses.

7. Describe the priming effect of a drug.

8. How do stressful situations provoke a relapse?

Lesson I Self Test

1. Which of the following is *not* true of drug use in general?

   a. Alcohol is one of the oldest drugs.
   b. Consciousness-changing drugs are sometimes used in religious ceremonies.
   c. Drugs have been cultivated around the world.
   d. Drug use is unique to modern Western cultures.

2. All of the following are linked to abuse of alcohol except:

   a. drug-induced psychosis.
   b. cirrhosis of the liver.
   c. mental retardation in infants exposed in utero.
   d. Korsakoff's syndrome.

3. Which statement about drug addiction is *not* true?

   a. Drug addiction is caused by physical dependence.
   b. Withdrawal symptoms and tolerance are the result of compensatory mechanisms.
   c. The withdrawal symptoms that occur in heroin addicts are less dangerous than those that occur in alcoholics.
   d. Withdrawal symptoms are primarily the opposite of the effects of the drug itself.

4. The most appropriate definition of drug addiction would emphasize

   a. negative personal, legal, and social effects of drug taking.
   b. physical dependence rather than psychological dependence.
   c. tolerance and withdrawal symptoms.
   d. an established pattern of compulsive drug-taking behavior.

5. Drugs that do not produce physical dependence

   a. can still be addictive.
   b. do not cause tolerance and withdrawal symptoms.
   c. have no reinforcing effects on behavior.
   d. have no effect on the brain.

6. All of the following are true of positive reinforcement except:

   a. Positive reinforcement increases the likelihood of the most recent response.
   b. A reinforcing stimulus is most effective if it occurs immediately after the behavior.
   c. Positive reinforcement works better in humans than in laboratory animals.
   d. Immediacy of reinforcement takes precedence over quantity.

7. Addictive drugs can trigger the release of dopamine in the nucleus accumbens by

   a. stimulating the reuptake of dopamine by terminal buttons.
   b. blocking the secretion of endogenous opioids.
   c. increasing the activity of dopaminergic neurons of the mesolimbic system.
   d. inhibiting the postsynaptic effects of dopamine.

8. Lamb et al. (1991) found that subjects with a history of heroin abuse

   a. push a lever very slowly.
   b. fail to push a lever for "worthless" doses of morphine.
   c. show that positive reinforcement requires pleasurable stimuli.
   d. show that positive reinforcement does not necessarily equal pleasure.

9. Negative reinforcement does *not* refer to

   a. a behavior that reduces an aversive stimulus.
   b. a response that makes an unpleasant stimulus occur.
   c. the capacity of alcohol to relieve feelings of anxiety.
   d. an addict suffering from withdrawal symptoms who takes some of the drug.

10. All of the following are compensatory mechanisms that have been proposed to explain the phenomena of tolerance to and withdrawal from drugs except

    a. downregulation of opiate receptors.
    b. reduced effects of opioid receptors on physiological processes within cells on which they are found.
    c. classical conditioning of homeostatic response to environmental stimuli.
    d. dopamine receptors in the nucleus accumbens become supersensitive.

11. What parts of the brain may be involved in a craving for addictive drugs?

a. dorsolateral prefrontal cortex, amygdala, and cerebellum
b. amygdala, periaqueductal gray matter
c. cerebellum, nucleus accumbens, reticular formation
d. nucleus accumbens, preoptic area, hippocampus

12. Which stimuli will probably *not* induce drug-seeking behavior?

a. the priming effect of a small dose
b. stressful situations
c. fatigue
d. the sight of drug-related paraphernalia

Lesson II: Commonly Abused Drugs, Heredity, and Therapy for Drug Abuse

Read the interim summary on pages 583-584 in your text to re-acquaint yourself with the material in this section.

Learning Objective 19-4   Review the neural basis of the reinforcing effects and withdrawal effects of opiates.

Read pages 571-575 and answer the following questions.

1.   Briefly discuss some of the serious personal and social costs of opiate addiction.

2.   To review: Why are endogenous opioids necessary for the survival of species?

3.   When opiate receptors are stimulated by an injection of an opiate, they produce different effects in the body. List four locations of opiate receptors and the effects they are responsible for.

1.   3.

2.   4.

4.   a. List the three major types of opiate receptors and the effects they produce when stimulated.

b. How did animals without mu opiate receptors respond to morphine? (Study Figure 19.5 in your text. Matthes et al., 1966)

c. How do kappa opiate receptors respond to dynorphin and other kappa receptor agonists? (Mucha and Herz, 1985; Suzuki et al., 1993)

5.   a. What is the general behavioral effect of injections of opiates into both ends of the mesolimbic dopaminergic system?

b. Why does this effect occur? (Johnson and North, 1992)

6. a. Explain how the apparatus shown in Figure 19.6 in your text was used in a conditioned place preference task by Olmstead and Franklin (1996)

b. When will an animal develop a conditioned place preference? a conditioned place aversion?

c. How did destruction of the nucleus accumbens affect the development of a conditioned place preference to amphetamine? to morphine?

d. What do the results suggest about the way opiates reinforce behavior?

7. Using the conditioned place preference procedure, what did Agmo et al., (1993) find about the physiology of reinforcement produced by natural reinforcers?

8. What regions of the brain may be involved in the withdrawal effects of opiates?

9. a. Define *antagonist-precipitated withdrawal* in your own words.

b. Using this technique, what did researchers discover about the effects of the sudden blocking of opiate receptors in particular brain regions? (Maldonado et all, 1992; Bozarth, 1994)

c. How do lesions of the locus coeruleus affect the severity of withdrawal symptoms? (Maldonado and Koob, 1993)

d. How does withdrawal affect the level of neurotransmitters in the locus coeruleus? (Aghajanian et al., 1994)

10 a. Identify a protein involved in the development of withdrawal symptoms.

b. What is its probable role? (Study Figure 19.7 in your text. Maldonado et al., 1996)

Learning Objective 19-5   Describe the behavioral and physical effects of cocaine, amphetamine, and nicotine.

Read pages 575-579 and answer the following questions.

1.   a. Why do cocaine and amphetamine have similar behavioral effects?

    b   What are their respective sites of action?

2.   Review how Giros et al., (1996) confirmed that the reinforcing effects of cocaine involve the dopamine transporter molecules.

    a. How did a targeted mutation affect the dopamine level in the brains of mice subjects?

    b. What kind of compensatory mechanisms did the researchers find in the brains of these mice?

    c. How did the compensatory mechanisms affect dopamine reuptake and the effects of cocaine and amphetamine?

    d. What do the results suggest about the role of dopamine transporters?

3.   a. How do people and laboratory animals behave after they have taken cocaine? (Geary, 1987)

    b. If rats are given continuous access to self-administered cocaine, what often happens? (Study Figure 19.8 in your text. Bozarth and Wise, 1985)

4.   Describe the psychotic behavior that usually results from the regular use of cocaine and amphetamine.

5.   Let's examine some of the effects on the brain of the intravenous injections of cocaine and amphetamine.

    a. How do injections of both drugs affect the nucleus accumbens? (Study Figure 19.9 in your text. Petit and Justice, 1989; Di Ciano et al, 1995; Wise et al., 1995)

    b. Which drug produces reinforcing effects? Which does not? (Hoebel et al., 1983; Phillips et al., 1994; Goeders and Smith, 1983)

    c. What is a possible explanation of these findings?

d. What happens to the behavioral effects of these drugs

    1. if drugs that block dopamine receptors are injected into the nucleus accumbens? (McGregor and Roberts, 1993; Caine et al., 1995)

    2. if lesions are produced in the nucleus accumbens?

    3. if 6-HD is injected there? (Study Figure 19.10 in your text. Caine and Koob, 1994)

6. Carefully explain how long-term use of cocaine or amphetamine may sensitize neurons in the nucleus accumbens. (Study Figure 19.11 in your text. Hyman et al., 1996a)

7. Summarize some of the health risks and explain the addictive potential of nicotine.

8. How does nicotine affect

a. acetylcholine receptors?

b. dopaminergic neurons of the mesolimbic system? (Mereu et al., 1987)

c. the nucleus accumbens? (Study Figure 19.12 in your text. Damsma et al., 1989)

9. Where does the reinforcing effect of nicotine appear to occur? Cite research to support your answer. (For example, study Figure 19.13 in your text. Nisell et al., 1994)

10. What are some of the symptoms of nicotine withdrawal? Explain why they contribute a resumption of smoking, but not to beginning to smoke.

11. Why may smoking and drinking alcohol make it more difficult to withdraw from cocaine or heroin use. Be sure to refer to the role of dopamine in reinforcement in your answer.

Learning Objective 19-6 Describe the behavioral and physical effects of alcohol, barbiturates, and cannabis.

Read pages 579-583 and answer the following questions.

1. To establish the importance of understanding the behavioral and physiological effects of this drug, summarize some of the social costs of alcohol use, especially its effect on brain development. (Be sure to look at Figure 19.14 in your text. Abel and Sokol, 1986)

2. What are the behavioral effects of low doses of alcohol? high doses of alcohol? Be sure to use the term *anxiolytic* in your answer.

3. Explain how the anxiolytic effects of alcohol influence the behavior of both laboratory animals and people. (Koob et al., 1984)

4. How can animals be induced to become dependent on alcohol? (Reid, 1996)

5. a Like other addictive drugs, what does alcohol do?

    b. In low to moderate doses, where are the major sites of action of alcohol?

6. a. Describe the *drug discrimination procedure.*

    b. What did Shelton and Balster's (1994) experiment conclude about the perceptual effects of alcohol?

7. a. Cite evidence that alcohol acts as an NMDA antagonist.

    b. Outline the sequence of events involved in withdrawal from long-term alcohol intake.

8. a. Compare alcohol withdrawal symptoms to heroin withdrawal symptoms.

    b. What kind of receptors may be responsible for the seizures that accompany alcohol withdrawal? (Valverius et al., 1990; Liljequist, 1991)

9. Describe how alcohol and the $GABA_A$ receptor interact.

10. Describe the "wonder drug" discovered by Suzdak et al. (1986) and why it is unlikely to be marketed. (See Figure 19.15 in your text.)

11. The effects of alcohol and barbiturates are _____ and the combined use of both drugs can be

    _____.

12. a. Name the active ingredient in marijuana and describe its effects on the release of dopamine. (See Figure 19.16 in your text. Chen et al., 1990)

    b. Outline a possible explanation for the memory deficits induced by marijuana.

    c. Summarize the research by Fletcher et al. (1995) on the long-term effects of cannabis use. (See Figure 19.17 in your text.)

Read the interim summary on page 587 in your text to re-acquaint yourself with the material in this section.

> *Learning Objective 19-7* Describe research on the role that heredity plays in addiction.

Read pages 584-587 and answer the following questions.

1. _____ and _____ are the only two possible sources of individual differences in any characteristic.

2. Explain why most research on the effects of hereditary on addiction has focused on alcoholism and why the results of these studies may apply to other forms of addiction as well.

3. Briefly describe characteristics linked to nicotine addiction.

4. a. Compare the concordance rate for alcoholism of monozygotic and dizygotic twins. (Goodwin, 1979)

    b. Circle the factor—heredity or environment—which is much more important in the development of alcoholism. (Cloninger et al., 1985)

5. Briefly describe the behavior of the two principal types of alcoholics. (See Table 19.1 in your text.)

    a. age at which drinking usually begins

b. drinking patterns

c. personal characteristics

6. a. How did having a steady drinking biological father affect the sons? the daughters? Be sure to use the term *somatization disorder* in your answer.

b. How does the family environment appear to influence the development of steady drinking alcoholism?

c. What circumstances contribute to binge drinking?

7. a. Briefly describe the social behavior of steady and binge drinkers. (Cloninger, 1987)

b. Describe the EEGs of binge drinkers. (Propping et al., 1981)

c. How do binge drinkers report they feel after taking alcohol? (Propping et al., 1980)

8. Explain how particular punishment or reinforcement systems may contribute to both forms of alcoholism.

9. Describe the discrepancy in recent research on differences in dopaminergic mechanisms and susceptibility to addiction. (Blum et al., 1990; Gelernter et al., 1993; Neiswanger et al., 1995; Noble et al., 1994)

10. Describe how the use of animal models has contributed to the understanding of alcohol addiction by comparing the behavior of alcohol preferring rats and alcohol nonpreferring rats when given

a. the choice of drinking or not drinking alcohol. (Li et al., 1993. See Figure 19.18 in your text.)

b. small doses and high doses of alcohol. (Gongwer et al., 1989; McBride et al., 1991)

c. What are some possible explanations for these behaviors? (For example, Li et al., 1993)

11. What did Higley et al. (1996) observe in rhesus monkeys with low levels of 5-HIAA in their cerebrospinal fluid?

Read the interim summary on pages 589-590 in your text to re-acquaint yourself with the material in this section.

Learning Objective 19-8   Discuss different methods of therapy for drug abuse.

Read pages 587-589 and answer the following questions.

1. Describe the most common treatment for opiate addiction.

2. a. What drugs are available in emergency rooms for patients who have taken an overdose of heroin?

   b. How can one of these drugs be used for heroin withdrawal? (Sandler and Freundlich, 1996)

3. a. Why are drugs that block dopamine receptors not useful treatments for cocaine and amphetamine addiction?

   b. How might D_3 receptor blockers be useful in treating cocaine dependency? (Staley and Mash, 1996)

4. Describe how Carrera et al. (1996) managed to "immunize" rats to cocaine.

5. Describe nicotine maintenance as a treatment for withdrawal from cigarette smoking. (See Figure 19.19 in your text.)

6. How do serotonin agonists help treat addiction to alcohol?

7. a. How do opiate antagonists alter the reinforcing effects of alcohol consumption?

   b. Describe the results of research on the effects of naltrexone on social drinking. (Davidson et al., 1996)

c. How effective is naltrexone in programs designed to treat alcohol abuse? (See Figure 19.20 in your text. O'Brien et al., 1996)

Lesson II Self Test

1. The reinforcing aspect of opiate drugs involves receptors in the

   a. periaqueductal gray matter.
   b. nucleus accumbens.
   c. preoptic area.
   d. reticular formation.

2. Which opiate receptors are responsible for the reinforcing effects of opiates?

   a. mu
   b. delta
   c. kappa
   d. alpha

3. Which of the following is *not* true of CREB?

   a. CREB stands for cyclic AMP-responsive enzyme-bonding protein.
   b. It is involved in the intracellular processes involved in the development of withdrawal symptoms of opiates.
   c. It plays a role in regulating the activity of some genes.
   d. It binds with a second messenger.

4. Cocaine

   a. causes strong physical dependence.
   b. is a dopamine antagonist.
   c. may be the most effective reinforcer of all available drugs.
   d. produces the same physical and behavioral effects as heroin.

5. Amphetamine

   a. causes dopamine transporters to run in reverse.
   b. stimulates the release of endogenous opiates.
   c. increases the sensitivity of GABA receptors.
   d. affects an enzyme in the second messenger system.

6. Which of the following is *not* true about nicotine?

   a. It has considerable addictive potential.
   b. It stimulates acetylcholine receptors.
   c. It is self-administered in the form of tobacco, but not by animals.

   d. It causes dopamine release in the nucleus accumbens.

7. Which of the following is *not* true of alcohol?

   a. It is the leading cause of mental retardation in children whose mothers consumed alcohol while pregnant.
   b. It results in sedation and incoordination at low doses.
   c. It produces both positive and negative reinforcement.
   d. Its anxiolytic effects are observed in both humans and animals.

8. The anxiolytic effect of alcohol

   a. reduces the discomfort of anxiety.
   b. forces the drinker to consume more and more to feel the same effects.
   c. reinforces social controls on behavior.
   d. provides positive reinforcement.

9. The effects of alcohol and barbiturates are *not*

   a. additive.
   b. sedative.
   c. excitatory.
   d. potentially fatal.

10. Marijuana may affect short-term memory by disrupting the normal function of the

    a. acetylcholine receptors.
    b. GABA receptors.
    c. nucleus accumbens.
    d. hippocampus.

11. Steady drinkers are most likely to

    a. be male; have a biological parent who is a steady drinker; begin drinking late in life.
    b. be male; have a father who is a steady drinker; begin drinking early in life.
    c. be female; have two biological parents who are alcoholic; drink secretly.
    d. be either male or female; be exposed to a family environment of heavy drinking; begin drinking early in life.

12. The most common treatment for opiate addiction is

a. naltrexone treatment.
b. treatment with drugs that block dopamine receptors.

c. treatment with serotonin agonists.
d. methadone maintenance.

Answers for Self Tests

Lesson I

1. d Obj. 19-1
2. a Obj. 19-1
3. a Obj. 19-1
4. d Obj. 19-1
5. a Obj. 19-1
6. c Obj. 19-2
7. c Obj. 19-2
8. d Obj. 19-2
9. b Obj. 19-2
10. d Obj. 19-3
11. a Obj. 19-3
12. c Obj. 19-3

Lesson II

1. b Obj. 19-4
2. a Obj. 19-4
3. a Obj. 19-4
4. c Obj. 19-5
5. a Obj. 19-5
6. c Obj. 19-5
7. b Obj. 19-6
8. a Obj. 19-6
9. c Obj. 19-6
10. d Obj. 19-6
11. b Obj. 19-7
12. d Obj. 19-8

1.1 dualism	1.10 model
1.2 monism (*mahn* ism)	1.11 doctrine of specific nerve energies
1.3 blindsight	1.12 experimental ablation
1.4 corpus callosum (*core* pus ka *low sum*)	1.13 functionalism
1.5 split-brain operation	1.14 natural selection
1.6 cerebral hemispheres	1.15 mutation
1.7 generalization	1.16 selective advantage
1.8 reduction	1.17 physiological psychologist
1.9 reflex	2.1 sensory neuron

1.10 A mathematical or physical analogy for a physiological process; for example, computers have been used as models for various functions of the brain.	1.1 The belief that the body is physical but the mind (or soul) is not.
1.11 Müller's conclusion that because all nerve fibers carry the same type of message, sensory information must be specified by the particular nerve fibers that are active.	1.2 The belief that the world consists only of matter and energy and the mind is part of it.
1.12 The research method in which the function of a part of the brain is inferred by observing the behaviors an animal can no longer perform after that part is damaged.	1.3 The ability of a person who cannot see objects in his or her blind field to accurately reach for them while remaining unconscious of perceiving them; caused by damage to the "mammalian" visual system of the brain.
1.13 The principle that the best way to understand a biological phenomenon (a behavior or a physiological structure) is to try to understand its useful functions for the organism.	1.4 The largest commissure of the brain, interconnecting the areas of neocortex on each side of the brain.
1.14 The process by which inherited traits that confer a selective advantage (increase an animal's likelihood to live and reproduce) become more prevalent in the population.	1.5 Brain surgery occasionally performed to treat a form of epilepsy; surgeon cuts the corpus callosum, which connects the two hemispheres of the brain.
1.15 A change in the genetic information contained in the chromosomes of gametes, which can be passed on to an organism's offspring; provides genetic variability.	1.6 The two symmetrical halves of the brain; constitute the major part of the brain.
1.16 A characteristic of an organism that permits it to produce more than the average number of offspring of its species.	1.7 Type of scientific explanation; a general conclusion based on many observations of similar phenomena.
1.17 A scientist who studies the physiology of behavior, primarily by performing physiological and behavioral experiments with laboratory animals.	1.8 Type of scientific explanation; a phenomenon is described in terms of the more elementary processes that underlie it.
2.1 A neuron that detects changes in the external or internal environment and sends information about these changes to the central nervous system.	1.9 An automatic, stereotyped movement produced as the direct result of a stimulus.

2.2 motor neuron	2.11 bipolar neuron
2.3 interneuron	2.12 unipolar neuron
2.4 central nervous system (CNS)	2.13 terminal button
2.5 peripheral nervous system (PNS)	2.14 transmitter substance/neurotransmitter
2.6 soma	2.15 membrane
2.7 dendrite	2.16 nucleus
2.8 synapse	2.17 nucleolus (*new clee o lus*)
2.9 axon	2.18 ribosome (*ry bo soam*)
2.10 multipolar neuron	2.19 chromosome

2.11 A neuron with one axon and one dendrite attached to its soma.	**2.2** A neuron located within the central nervous system that controls the contraction of a muscle or the secretion of a gland.
2.12 A neuron with one axon attached to its soma; the axon divides, with one branch receiving sensory information and the other sending the information into the central nervous system.	**2.3** A neuron located entirely within the central nervous system.
2.13 The bud at the end of a branch of an axon; forms synapses with another neuron; sends information to that neuron.	**2.4** The brain and spinal cord.
2.14 A chemical that is released by a terminal button; has an excitatory or inhibitory effect on another neuron.	**2.5** That part of the nervous system outside the brain and spinal cord, including the nerves attached to the brain and spinal cord.
2.15 A structure consisting principally of lipid molecules that defines the outer boundaries of a cell and also constitutes many of the cell organelles, such as the Golgi apparatus.	**2.6** The cell body of a neuron, which contains the nucleus.
2.16 A structure in the central region of a cell, enclosed by a membrane, containing the nucleolus and chromosomes.	**2.7** A branched, treelike structure attached to the soma of a neuron; receives information from the terminal buttons of other neurons.
2.17 An structure within the nucleus of a cell that produces the ribosomes.	**2.8** A junction between the terminal button of an axon and the membrane of another neuron.
2.18 A cytoplasmic structure, made of protein, that serves as the site of production of proteins translated from mRNA.	**2.9** The long, thin, cylindrical structure that conveys information from the soma of a neuron to its terminal buttons.
2.19 A strand of DNA, with associated proteins, found in the nucleus; carries genetic information.	**2.10** A neuron with one axon and many dendrites attached to its soma.

2.20 deoxyribonucleic acid (DNA) (*dee ox ee ry bo new clay ik*)	2.29 exocytosis (*ex o sy toe sis*)
2.21 gene	2.30 lysosome (*lye so soam*)
2.22 messenger ribonucleic acid (mRNA)	2.31 cytoskeleton
2.23 enzyme	2.32 microtubule (*my kro too bule*)
2.24 cytoplasm	2.33 neurofilament
2.25 mitochondria	2.34 microfilament
2.26 adenosine triphosphate (ATP) (*ah den o seen*)	2.35 axoplasmic transport
2.27 endoplasmic reticulum	2.36 anterograde
2.28 Golgi apparatus (*goal jee*)	2.37 retrograde

2.29 The secretion of a substance by a cell through means of vesicles; the process by which neurotransmitters are secreted.	2.20 A long, complex macromolecule consisting of two interconnected helical strands; along with associated proteins, strands of DNA constitute the chromosomes.
2.30 An organelle surrounded by membrane; contains enzymes that break down waste products.	2.21 The functional unit of the chromosome, which directs synthesis of one or more proteins.
2.31 Formed of microtubules, neurofilaments, and microfilaments, linked to each other and forming a cohesive mass that gives a cell its shape.	2.22 A macromolecule that delivers genetic information concerning the synthesis of a protein from a portion of a chromosome to a ribosome.
2.32 A long strand of bundles of protein filaments arranged around a hollow core; part of the cytoskeleton and involved in transporting substances from place to place within the cell.	2.23 A molecule that controls a chemical reaction, combining two substances or breaking a substance into two parts.
2.33 One of the fibers of the cytoskeleton, made of long, continuous strands of protein similar to those found in hair.	2.24 The viscous, semiliquid substance contained in the interior of a cell.
2.34 The thinnest of the fibers of the cytoskeleton; forms a meshwork just inside the membrane that holds membrane-bound proteins in place.	2.25 An organelle responsible for extracting energy from nutrients.
2.35 An active process by which substances are propelled along microtubules that run the length of the axon.	2.26 A molecule of prime importance to cellular energy metabolism; its breakdown liberates energy.
2.36 In a direction along an axon from the cell body toward the terminal buttons.	2.27 Parallel layers of membrane found within the cytoplasm of a cell. Rough endoplasmic reticulum contains ribosomes; smooth endoplasmic reticulum is the site of synthesis of lipids and provides channels for the segregation of molecules involved in various cellular processes.
2.37 In a direction along an axon from the terminal buttons toward the cell body.	2.28 A complex of parallel membranes in the cytoplasm that wraps the products of a secretory cell.

2.38 glia (*glee* ah)	2.47 area postrema (*poss* **tree** *ma*)
2.39 astrocyte	2.48 electrode/microelectrode
2.40 phagocytosis (**fagg** *o sy* **toe** *sis*)	2.49 membrane potential
2.41 oligodendrocyte (*oh li go* **den** *droh site*)	2.50 oscilloscope
2.42 myelin sheath (*my a lin*)	2.51 resting potential
2.43 node of Ranvier (**raw** *vee ay*)	2.52 depolarization
2.44 microglia	2.53 hyperpolarization
2.45 Schwann cell	2.54 action potential
2.46 blood-brain barrier	2.55 threshold of excitation

2.47 A region of the medulla where the blood-brain barrier is weak; poisons can be detected there and can initiate vomiting.	**2.38** The supporting cells of the central nervous system.
2.48 An electrode is a conductive medium that can be used to apply electrical stimulation or to record electrical potentials. A microelectrode is a very fine electrode, generally used to record activity of individual neurons.	**2.39** A glial cell that provides support for neurons of the central nervous system, provides nutrients and other substances, and regulates the chemical composition of the extracellular fluid.
2.49 The electrical charge across a cell membrane; the difference in electrical potential inside and outside the cell.	**2.40** The process by which cells engulf and digest other cells or debris caused by cellular degeneration.
2.50 A laboratory instrument capable of displaying a graph of voltage as a function of time on the face of a cathode ray tube.	**2.41** A type of glial cell in the central nervous system that forms myelin sheaths.
2.51 The membrane potential of a neuron when it is not being altered by excitatory or inhibitory postsynaptic potentials; approximately -70 mV in the giant squid axon.	**2.42** A sheath that surrounds axons and insulates them, preventing messages from spreading between adjacent axons.
2.52 Reduction (toward zero) of the membrane potential of a cell from its normal resting potential.	**2.43** A naked portion of a myelinated axon, between adjacent oligodendroglia or Schwann cells.
2.53 An increase in the membrane potential of a cell, relative to the normal resting potential.	**2.44** The smallest of glial cells; act as phagocytes and protect the brain from invading microorganisms.
2.54 The brief electrical impulse that provides the basis for conduction of information along an axon.	**2.45** A cell in the peripheral nervous system that is wrapped around a myelinated axon, providing one segment of its myelin sheath.
2.55 The value of the membrane potential that must be reached in order to produce an action potential.	**2.46** A semipermeable barrier produced by the astrocytes and cells in the walls of the capillaries in the brain.

2.56 diffusion	2.65 all-or-none law
2.57 electrolyte	2.66 rate law
2.58 ion	2.67 cable properties
2.59 electrostatic pressure	2.68 saltatory conduction
2.60 intracellular fluid	2.69 postsynaptic potential
2.61 extracellular fluid	2.70 neuromodulator
2.62 sodium-potassium transporter	2.71 endocrine gland
2.63 ion channel	2.72 target cell
2.64 voltage-dependent ion channel	2.73 binding site

2.65 The principle that once an action potential is triggered in an axon, it is propagated, without decrement, to the end of the fiber.	2.56 Movement of molecules from regions of high concentration to regions of low concentration.
2.66 The principle that variations in the intensity of a stimulus or other information being transmitted in an axon are represented by variations in the rate at which that axon fires.	2.57 An aqueous solution of a material that ionizes-namely, a soluble acid, base, or salt.
2.67 The passive conduction of electrical current, in a decremental fashion, down the length of an axon.	2.58 A charged molecule. *Cations* are positively charged, and *anions* are negatively charged.
2.68 Conduction of action potentials by myelinated axons. The action potential "jumps" from one node of Ranvier to the next.	2.59 The attractive force between atomic particles charged with opposite signs, or the repulsive force between atomic particles charged with the same sign.
2.69 Alterations in the membrane potential of a postsynaptic neuron, produced by liberation of transmitter substance at the synapse.	2.60 The fluid contained within cells.
2.70 A naturally secreted substance that acts like a neurotransmitter except that it is not restricted to the synaptic cleft but diffuses through the interstitial fluid.	2.61 Body fluids located outside of cells.
2.71 A gland that liberates its secretions into the extracellular fluid around capillaries and hence into the bloodstream.	2.62 A protein found in the membrane of all cells that extrudes sodium ions from and transports potassium ions into the cell.
2.72 The type of cell that is directly affected by a hormone or nerve fiber.	2.63 A specialized protein molecule that permits specific ions to enter or leave cells.
2.73 The location on a receptor protein to which a ligand binds.	2.64 An ion channel that opens or closes according to the value of the membrane potential.

2.74 ligand (**ligh** gand or **ligg** and)	2.83 neurotransmitter-dependent ion channel
2.75 dendritic spine	2.84 ionotropic receptor (*eye on oh* **trow** *pik*)
2.76 presynaptic/postsynaptic membrane	2.85 metabotropic receptor (*meh tab oh* **trow** *pik*)
2.77 synaptic cleft	2.86 G protein
2.78 synaptic vesicle (**vess** *i kul*)	2.87 second messenger
2.79 release zone	2.88 reuptake
2.80 cisterna	2.89 enzymatic deactivation
2.81 pinocytosis (*pee no sy* **toh** *sis*)	2.90 acetylcholine (ACh) (*a see tul* **koh** *leen*)
2.82 postsynaptic receptor	2.91 acetylcholinesterase (AChE) (*a see tul koh lin* **ess** *ter ace*)

2.83 An ion channel that opens when a molecule of a neurotransmitter binds with a postsynaptic receptor.	**2.74** A chemical that binds with the binding site of a receptor.
2.84 A receptor that contains a binding site for a neurotransmitter and an ion channel that opens when a molecule of the neurotransmitter attaches to the binding site.	**2.75** A small bud on the surface of a dendrite, with which a terminal button from another neuron forms a synapse.
2.85 A receptor that contains a binding site for a neurotransmitter; activates an enzyme that begins a series of events that opens an ion channel elsewhere in the membrane of the cell when a molecule of the neurotransmitter attaches to the binding site.	**2.76** *Presynaptic:* The membrane of a terminal button that lies adjacent to the postsynaptic membrane. *Postsynaptic* The cell membrane opposite the terminal button in a synapse; the membrane of the cell that receives the message.
2.86 A protein coupled to a metabotropic receptor; conveys messages to other molecules when a ligand binds with and activates the receptor.	**2.77** The space between the presynaptic membrane and the postsynaptic membrane.
2.87 A chemical produced when a G protein activates an enzyme; carries a signal that results in the opening of the ion channel or causes other events to occur in the cell.	**2.78** A small, hollow, beadlike structure found in terminal buttons; contains molecules of a neurotransmitter.
2.88 The reentry of a transmitter substance just liberated by a terminal button back through its membrane, thus terminating the postsynaptic potential.	**2.79** A region of the interior of the postsynaptic membrane of a synapse to which synaptic vesicles attach and release their neurotransmitter into the synaptic cleft.
2.89 The destruction of a transmitter substance by an enzyme after its release-for example, the destruction of acetylcholine by acetylcholinesterase.	**2.80** A part of the Golgi apparatus; through the process of pinocytosis, it receives portions of the presynaptic membrane and recycles them into synaptic vesicles.
2.90 A neurotransmitter found in the brain, spinal cord, and parts of the peripheral nervous system; responsible for muscular contraction.	**2.81** The pinching off of a bud of cell membrane, which travels to the interior of the cell.
2.91 The enzyme that destroys acetylcholine soon after it is liberated by the terminal buttons, thus terminating the postsynaptic potential.	**2.82** A receptor molecule in the postsynaptic membrane of a synapse that contains a binding site for a neurotransmitter.

2.92 neural integration	3.4 dorsal/ventral
2.93 autoreceptor	3.5 lateral
2.94 presynaptic inhibition/facilitation	3.6 medial
2.95 gap junction	3.7 ipsilateral/contralateral
2.96 peptide	3.8 cross section/frontal section
2.97 steroid	3.9 horizontal section
3.1 neuraxis	3.10 sagittal section (*sadj i tul*)
3.2 anterior/posterior	3.11 midsagittal plane
3.3 rostral/caudal	3.12 central nervous system (**CNS**)

3.4 "Toward the back/toward the belly"; with respect to the central nervous system, in a direction perpendicular to the neuraxis toward the top of the head or the back/toward the bottom of the skull or the front surface of the body.	**2.92** The process by which inhibitory and excitatory postsynaptic potentials summate and control the rate of firing of a neuron.
3.5 Toward the side of the body, in a direction at right angles with the neuraxis and away from it.	**2.93** A receptor molecule located on a neuron that responds to the neurotransmitter released by that neuron.
3.6 Toward the neuraxis, away from the side of the body.	**2.94** The action of a presynaptic terminal button in an axoaxonic synapse; reduces/increases the amount of neurotransmitter released by the postsynaptic terminal button.
3.7 Located on the same side of the body/the opposite side of the body.	**2.95** A special junction between cells that permits direct communication by means of electrical coupling.
3.8 With respect to the central nervous system, a slice taken at right angles to the neuraxis.	**2.96** A chain of amino acids joined together by peptide bonds.
3.9 A slice through the brain parallel to the ground.	**2.97** A chemical of low molecular weight, derived from cholesterol. Steroid hormones affect their target cells by attaching to receptors found within the cell.
3.10 A slice through the brain parallel to the neuraxis and perpendicular to the ground.	**3.1** An imaginary line drawn through the center of the length of the central nervous system, from the bottom of the spinal cord to the front of the forebrain.
3.11 The plane through the neuraxis perpendicular to the ground; divides the brain into two symmetrical halves.	**3.2** With respect to the central nervous system, located near or toward the head/tail.
3.12 The brain and spinal cord.	**3.3** "Toward the beak/toward the tail"; with respect to the central nervous system, in a direction along the neuraxis toward/away from the front of the face.

3.13 peripheral nervous system (**PNS**)	3.22 ventricle (***ven trik ul***)
3.14 vertebral artery (*ver **tee** brul*)	3.23 lateral ventricle
3.15 internal carotid artery	3.24 third ventricle
3.16 meninges (singular: meninx) (*men **in** jees*)	3.25 cerebral aqueduct
3.17 dura mater	3.26 fourth ventricle
3.18 arachnoid membrane (*a **rak** noyd*)	3.27 choroid plexus
3.19 pia mater	3.28 arachnoid granulation
3.20 subarachnoid space	3.29 superior sagittal sinus
3.21 cerebrospinal fluid (**CSF**)	3.30 obstructive hydrocephalus

3.22 One of the hollow spaces within the brain, filled with cerebrospinal fluid.	3.13 The nerves and ganglia located outside the central nervous system.
3.23 One of the two ventricles located in the center of the telencephalon.	3.14 An artery whose branches serve the posterior region of the brain.
3.24 The ventricle located in the center of the diencephalon.	3.15 An artery whose branches serve the rostral and lateral portions of the brain.
3.25 A narrow tube interconnecting the third and fourth ventricles of the brain, located in the center of the mesencephalon.	3.16 The three layers of tissue that encase the central nervous system: the dura mater, arachnoid membrane, and pia mater.
3.26 The ventricle located between the cerebellum and the dorsal pons, in the center of the metencephalon.	3.17 The outermost of the meninges; tough, flexible, unstretchable.
3.27 The highly vascular tissue that protrudes into the ventricles and produces cerebrospinal fluid.	3.18 The middle layer of the meninges, between the outer dura mater and inner pia mater. The subarachnoid space beneath the arachnoid membrane is filled with cerebrospinal fluid, which cushions the brain.
3.28 Small projections of the arachnoid membrane through the dura mater into the superior sagittal sinus; CSF flows through them to be reabsorbed into the blood supply.	3.19 The layer of the meninges adjacent to the surface of the brain.
3.29 A venous sinus located in the midline just dorsal to the brain, between the two cerebral hemispheres.	3.20 The fluid-filled space between the arachnoid membrane and the pia mater.
3.30 A condition in which all or some of the brain's ventricles are enlarged; caused by an obstruction that impedes the normal flow of CSF.	3.21 A clear fluid, similar to blood plasma, that fills the ventricular system of the brain and the subarachnoid space surrounding the brain and spinal cord.

3.31 forebrain	3.40 primary auditory cortex
3.32 cerebral hemisphere (*sa **ree** brul*)	3.41 lateral fissure
3.33 subcortical region	3.42 primary somatosensory cortex
3.34 cerebral cortex	3.43 central sulcus (***sul** kus*)
3.35 sulcus (plural: sulci) (***sul** kus, **sul** sigh*)	3.44 primary motor cortex
3.36 fissure	3.45 frontal lobe
3.37 gyrus (plural: gyri) (***jye** russ, **jye** rye*)	3.46 parietal lobe (*pa **rye** i tul*)
3.38 primary visual cortex	3.47 temporal lobe (***tem** por ul*)
3.39 calcarine fissure (***kal** ka rine*)	3.48 occipital lobe (*ok **sip** i tul*)

3.40 The region of the cerebral cortex whose primary input is from the auditory system.	3.31 The most rostral of the three major divisions of the brain; includes the telencephalon and diencephalon.
3.41 The fissure that separates the temporal lobe from the overlying frontal and parietal lobes.	3.32 One of the two major portions of the forebrain, covered by the cerebral cortex.
3.42 The region of the cerebral cortex whose primary input is from the somatosensory system.	3.33 The region located within the brain, beneath the cortical surface.
3.43 The sulcus that separates the frontal lobe from the parietal lobe.	3.34 The outermost layer of gray matter of the cerebral hemispheres.
3.44 The region of the cerebral cortex that contains neurons that control movements of skeletal muscles.	3.35 A groove in the surface of the cerebral hemisphere, smaller than a fissure.
3.45 The anterior portion of the cerebral cortex, rostral to the parietal lobe and dorsal to the temporal lobe.	3.36 A major groove in the surface of the brain, larger than a sulcus.
3.46 The region of the cerebral cortex caudal to the frontal lobe and dorsal to the temporal lobe.	3.37 A convolution of the cortex of the cerebral hemispheres, separated by sulci or fissures.
3.47 The region of the cerebral cortex rostral to the occipital lobe and ventral to the parietal and frontal lobes.	3.38 The region of the cerebral cortex whose primary input is from the visual system.
3.48 The region of the cerebral cortex caudal to the parietal and temporal lobes.	3.39 A fissure located in the occipital lobe on the medial surface of the brain; contains most of the primary visual cortex.

3.49
 	
           sensory association cortex
	3.58
	
	           hippocampus

3.50
 	
           motor association cortex
	3.59
	
	           amygdala
	           (*a **mig** da la*)

3.51
 	
           prefrontal cortex
	3.60
	
	           fornix

3.52
 	
           corpus callosum
	3.61
	
	           mammillary bodies
	           (***mam** i lair ee*)

3.53
 	
           neocortex
	3.62
	
	           basal ganglia

3.54
 	
           limbic cortex
	3.63
	
	           diencephalon
	           (*dy en **seff** a lahn*)

3.55
 	
           cingulate gyrus
           (***sing** yew lett*)
	3.64
	
	           thalamus

3.56
 	
           commissure
           (***kahm** i sher*)
	3.65
	
	           projection fiber

3.57
 	
           limbic system
	3.66
	
	           nucleus
	           (plural: nuclei)

3.58 A forebrain structure of the temporal lobe, constituting an important part of the limbic system; includes the hippocampus proper (Ammon's horn), dentate gyrus, and subiculum.	3.49 Those regions of the cerebral cortex that receive information from the regions of primary sensory cortex.
3.59 A structure in the interior of the rostral temporal lobe, containing a set of nuclei; part of the limbic system.	3.50 The region of the frontal lobe rostral to the primary motor cortex.
3.60 A fiber bundle that connects the hippocampus with other parts of the brain, including the mammillary bodies of the hypothalamus.	3.51 The region of the frontal lobe rostral to the motor association cortex.
3.61 A protrusion of the bottom of the brain at the posterior end of the hypothalamus, containing some hypothalamic nuclei.	3.52 The largest commissure of the brain, interconnecting the areas of neocortex on each side of the brain.
3.62 A group of subcortical nuclei in the telencephalon, the caudate nucleus, the globus pallidus, and the putamen; important parts of the motor system.	3.53 The phylogenetically newest cortex, including the primary sensory cortex, primary motor cortex, and association cortex.
3.63 A region of the forebrain surrounding the third ventricle; includes the thalamus and the hypothalamus.	3.54 Phylogenetically old cortex, located at the edge ("limbus") of the cerebral hemispheres; part of the limbic system.
3.64 The largest portion of the diencephalon, located above the hypothalamus; contains nuclei that project information to specific regions of the cerebral cortex and receive information from it.	3.55 A strip of limbic cortex lying along the lateral walls of the groove separating the cerebral hemispheres, just above the corpus callosum.
3.65 An efferent axon from a neuron in one region of the brain whose terminals form synapses with neurons in another region.	3.56 A fiber bundle that interconnects corresponding regions on each side of the brain.
3.66 An identifiable group of neural cell bodies in the central nervous system.	3.57 A group of brain regions including the anterior thalamic nuclei, amygdala, hippocampus, limbic cortex, and parts of the hypothalamus, as well as their interconnecting fiber bundles.

3.67 lateral/medial geniculate nucleus	3.76 superior/inferior colliculi (*ka **lik** yew lee*)
3.68 ventrolateral nucleus	3.77 brain stem
3.69 hypothalamus	3.78 tegmentum
3.70 optic chiasm (***kye** az'm*)	3.79 reticular formation
3.71 anterior pituitary gland	3.80 periaqueductal gray matter
3.72 neurosecretory cell	3.81 red nucleus
3.73 posterior pituitary gland	3.82 substantia nigra
3.74 midbrain/mesencephalon (*mezz en **seff** a lahn*)	3.83 hindbrain
3.75 tectum	3.84 cerebellum (*sair a **bell** um*)

3.76 Protrusions on top of the midbrain; part of the visual/auditory system.	3.67 A group of cell bodies within the lateral/medial geniculate body of the thalamus that receives fibers from the retina/auditory system and projects fibers to the primary visual/auditory cortex.
3.77 The "stem" of the brain, from the medulla to the diencephalon, excluding the cerebellum.	3.68 A nucleus of the thalamus that receives inputs from the cerebellum and sends axons to the primary motor cortex.
3.78 The ventral part of the midbrain; includes the periaqueductal gray matter, reticular formation, red nucleus, and substantia nigra.	3.69 The group of nuclei of the diencephalon situated beneath the thalamus; involved in regulation of the autonomic nervous system, control of the anterior and posterior pituitary glands, and integration of species-typical behaviors.
3.79 A large network of neural tissue located in the central region of the brain stem, from the medulla to the diencephalon.	3.70 A cross-shaped connection between the optic nerves, located below the base of the brain, just anterior to the pituitary gland.
3.80 The region of the midbrain surrounding the cerebral aqueduct; contains neural circuits involved in species-typical behaviors.	3.71 The anterior part of the pituitary gland; an endocrine gland whose secretions are controlled by the hypothalamic hormones.
3.81 A large nucleus of the midbrain that receives inputs from the cerebellum and motor cortex and sends axons to motor neurons in the spinal cord.	3.72 A neuron that secretes a hormone or hormonelike substance.
3.82 A darkly stained region of the tegmentum that contains neurons that communicate with the caudate nucleus and putamen in the basal ganglia.	3.73 The posterior part of the pituitary gland; an endocrine gland that contains hormone-secreting terminal buttons of axons whose cell bodies lie within the hypothalamus.
3.83 The most caudal of the three major divisions of the brain; includes the metencephalon and myelencephalon.	3.74 The central of the three major divisions of the brain, surrounds the cerebral aqueduct; includes the tectum and the tegmentum.
3.84 A major part of the brain located dorsal to the pons, containing the two cerebellar hemispheres, covered with the cerebellar cortex; an important component of the motor system.	3.75 The dorsal part of the midbrain; includes the superior and inferior colliculi.

3.85 cerebellar cortex	3.94 dorsal root/ventral root
3.86 deep cerebellar nuclei	3.95 spinal nerve
3.87 cerebellar peduncle (*pee dun kul*)	3.96 afferent axon/efferent axon
3.88 pons	3.97 dorsal root ganglion
3.89 medulla oblongata (*me doo la*)	3.98 cranial nerve
3.90 spinal cord	3.99 vagus nerve
3.91 spinal root	3.100 olfactory bulb
3.92 cauda equina (*ee kwye na*)	3.101 somatic nervous system
3.93 caudal block	3.102 autonomic nervous system (ANS)

3.94 The spinal root that contains incoming (afferent) sensory fibers/outgoing (efferent) motor fibers.	3.85 The cortex that covers the surface of the cerebellum
3.95 A peripheral nerve attached to the spinal cord.	3.86 Nuclei located within the cerebellar hemispheres; receive projections from the cerebellar cortex and send projections out of the cerebellum to other parts of the brain.
3.96 An axon directed toward/away from the central nervous system, conveying sensory information/motor commands.	3.87 One of three bundles of axons that attach each cerebellar hemisphere to the dorsal pons.
3.97 A nodule on a dorsal root that contains cell bodies of afferent spinal nerve neurons.	3.88 The region of the metencephalon rostral to the medulla, caudal to the midbrain, and ventral to the cerebellum.
3.98 A peripheral nerve attached directly to the brain.	3.89 The most caudal portion of the brain; located in the myelencephalon, immediately rostral to the spinal cord.
3.99 The largest of the cranial nerves, conveying efferent fibers of the parasympathetic division of the autonomic nervous system to organs of the thoracic and abdominal cavities.	3.90 The cord of nervous tissue that extends caudally from the medulla
3.100 The protrusion at the end of the olfactory nerve; receives input from the olfactory receptors.	3.91 A bundle of axons surrounded by connective tissue that occurs in pairs, which fuse and form a spinal nerve.
3.101 The part of the peripheral nervous system that controls the movement of skeletal muscles or transmits somatosensory information to the central nervous system.	3.92 A bundle of spinal roots located caudal to the end of the spinal cord.
3.102 The portion of the peripheral nervous system that controls the body's vegetative function.	3.93 The anesthesia and paralysis of the lower part of the body produced by injection of a local anesthetic into the cerebrospinal fluid surrounding the cauda equina.

3.103 sympathetic division	4.3 sites of action
3.104 spinal sympathetic ganglia	4.4 pharmacokinetics
3.105 sympathetic ganglion chain	4.5 intravenous (IV) injection
3.106 preganglionic neuron	4.6 intraperitoneal (IP) injection (*in tra pair i toe **nee** ul*)
3.107 postganglionic neuron	4.7 intramuscular (IM) injection
3.108 adrenal medulla	4.8 subcutaneous (SC) injection
3.109 parasympathetic division	4.9 oral administration
4.1 psychopharmacology	4.10 sublingual administration (*sub **ling** wul*)
4.2 drug effect	4.11 intrarectal administration

4.3 The locations at which molecules of drugs interact with molecules located on or in cells of the body, thus affecting some biochemical processes of these cells.	**3.103** The portion of the autonomic nervous system that controls functions that accompany arousal and expenditure of energy.
4.4 The process by which drugs are absorbed, distributed within the body, metabolized, and excreted.	**3.104** Sympathetic ganglia either adjacent to the spinal cord in the sympathetic chain or located in the abdominal cavity.
4.5 Injection of a substance directly into a vein.	**3.105** One of a pair of groups of sympathetic ganglia that lie ventrolateral to the vertebral column.
4.6 Injection of a substance into the *peritoneal cavity* the space that surrounds the stomach, intestines, liver, and other abdominal organs.	**3.106** The efferent neuron of the autonomic nervous system whose cell body is located in a cranial nerve nucleus or in the intermediate horn of the spinal gray matter and whose terminal buttons synapse upon postganglionic neurons in the autonomic ganglia.
4.7 Injection of a substance into a muscle.	**3.107** Neurons of the autonomic nervous system that form synapses directly with their target organ.
4.8 Injection of a substance into the space beneath the skin.	**3.108** The inner portion of the adrenal gland, located atop the kidney, controlled by sympathetic nerve fibers; secretes epinephrine and norepinephrine.
4.9 Administration of a substance into the mouth, so that it is swallowed.	**3.109** The portion of the autonomic nervous system that controls functions that occur during a relaxed state.
4.10 Administration of a substance by placing it beneath the tongue.	**4.1** The study of the effects of drugs on the nervous system and on behavior.
4.11 Administration of a substance into the rectum.	**4.2** The changes a drug produces in an organism's physiological processes and behavior.

4.12 inhalation	4.21 tolerance
4.13 topical administration	4.22 sensitization
4.14 intracerebral administration	4.23 withdrawal symptom
4.15 intracerebroventricular (ICV) administration	4.24 placebo (*pla **see** boh*)
4.16 depot binding	4.25 antagonist
4.17 albumin (*al **bew** min*)	4.26 agonist
4.18 dose-response curve	4.27 direct agonist
4.19 therapeutic index	4.28 receptor blocker
4.20 affinity	4.29 direct antagonist

4.21 A decrease in the effectiveness of a drug that is administered repeatedly.	4.12 Administration of a vaporous substance into the lungs.
4.22 An increase in the effectiveness of a drug that is administered repeatedly.	4.13 Administration of a substance directly onto the skin or mucous membrane.
4.23 The appearance of symptoms opposite to those produced by a drug when the drug is administered repeatedly and then suddenly no longer taken.	4.14 Administration of a substance directly into the brain.
4.24 An inert substance given to an organism in lieu of a physiologically active drug; used experimentally to control for the effects of mere administration of a drug.	4.15 Administration of a substance into one of the cerebral ventricles.
4.25 A drug that opposes or inhibits the effects of a particular neurotransmitter on the postsynaptic cell.	4.16 Binding of a drug with various tissues of the body or with proteins in the blood.
4.26 A drug that facilitates the effects of a particular neurotransmitter on the postsynaptic cell.	4.17 A protein found in the blood; serves to transport free fatty acids and can bind with some lipid-soluble drugs.
4.27 A drug that binds with and activates a receptor.	4.18 A graph of the magnitude of an effect of a drug as a function of the amount drug administered.
4.28 A drug that binds with a receptor but does not activate it ; prevents the natural ligand from binding with the receptor.	4.19 The ratio between the dose that produces the desired effect in 50 percent of the animals and the dose that produces toxic effects in 50 percent of the animals.
4.29 A synonym for receptor blocker.	4.20 The readiness with which two molecules join together.

4.30 noncompetitive binding	4.39 hemicholinium (*hem ee koh* **lin** *um*)
4.31 inverse agonist	4.40 nicotinic receptor
4.32 indirect agonist	4.41 muscarinic receptor (*muss ka* **rin** *ic*)
4.33 presynaptic heteroreceptor	4.42 atropine (***a*** *tro peen*)
4.34 acetyl-CoA (*a* ***see*** *tul*)	4.43 curare (*kew* **rahr** *ee*)
4.35 choline acetyltransferase (ChAT) (**koh** *leen a see tul* **trans** *fer ace*)	4.44 monoamine (**mahn** *o a meen*)
4.36 botulinum toxin (*bot you* **lin** *um*)	4.45 catecholamine (*cat a* **kohl** *a meen*)
4.37 black widow spider venom	4.46 dopamine (DA) (**dope** *a meen*)
4.38 neostigmine (*nee o* **stig** *meen*)	4.47 L-DOPA (*ell* **dope** *a*)

4.39 A drug that inhibits the uptake of choline.	**4.30** Binding of a drug to a site on a receptor; does not interfere with the binding site for the principal ligand.
4.40 An ionotropic acetylcholine receptor that is stimulated by nicotine and blocked by curare.	**4.31** A drug that attaches to a binding site on a receptor and interferes with the action of the receptor; does not interfere with the binding site for the principal ligand.
4.41 A metabotropic acetylcholine receptor that is stimulated by muscarine and blocked by atropine.	**4.32** A drug that attaches to a binding site on a receptor and facilitates the action of the receptor; does not interfere with the binding site for the principal ligand.
4.42 A drug that blocks muscarinic acetylcholine receptors.	**4.33** A receptor located in the membrane of a terminal button that receives input from another terminal button by means of an axoaxonic synapse; binds with the neurotransmitter released by the presynaptic terminal button.
4.43 A drug that blocks nicotinic acetylcholine receptors.	**4.34** A cofactor that supplies acetate for the synthesis of acetylcholine.
4.44 A class of amines that includes indolamines such as serotonin and catecholamines such as dopamine norepinephrine, and epinephrine.	**4.35** The enzyme that transfers the acetate ion from acetyl coenzyme A to choline, producing the neurotransmitter acetylcholine.
4.45 A class of amines that includes the neurotransmitters dopamine, norepinephrine, and epinephrine.	**4.36** An acetylcholine antagonist: prevents release by terminal buttons.
4.46 A neurotransmitter; one of the catecholamines.	**4.37** A poison produced by the black widow spider that triggers the release of acetylcholine.
4.47 The levorotatory form of DOPA; the precursor of the catecholamines; often used to treat Parkinson's disease because of its effect as a dopamine agonist.	**4.38** A drug that inhibits the activity of acetylcholinesterase.

4.48 nigrostriatal system (*nigh grow stry **ay** tul*)	4.57 deprenyl (***depp** ra nil*)
4.49 mesolimbic system (*mee zo **lim** bik*)	4.58 chlorpromazine (*klor **proh** ma zeen*)
4.50 mesocortical system (*mee zo **kor** ti kul*)	4.59 clozapine (***kloz** a peen*)
4.51 Parkinson's disease	4.60 norepinephrine (NE) (*nor epp i **neff** rin*)
4.52 AMPT	4.61 epinephrine (*epp i **neff** rin*)
4.53 reserpine (*ree **sur** peen*)	4.62 fusaric acid (*few **sahr** ik*)
4.54 apomorphine (*ap o **more** feen*)	4.63 moclobemide (*mak low **bem** ide*)
4.55 methylphenidate (*meth ul **fen** i date*)	4.64 locus coeruleus (*sur **oo** lee us*)
4.56 monoamine oxidase (MAO) (***mahn** o a meen*)	4.65 axonal varicosities

4.57 A drug that blocks the activity of MAO-B; acts as a dopamine agonist.	4.48 A system of neurons originating in the substantia nigra and terminating in the neostriatum (caudate nucleus and putamen).
4.58 A drug that reduces the symptoms of schizophrenia by blocking dopamine D_2 receptors.	4.49 A system of dopaminergic neurons originating in the ventral tegmental area and terminating in the nucleus accumbens, amygdala, and hippocampus.
4.59 A drug that reduces the symptoms of schizophrenia, apparently by blocking dopamine D_4 receptors.	4.50 A system of dopaminergic neurons originating in the ventral tegmental area and terminating in the prefrontal cortex.
4.60 One of the catecholamines; a neurotransmitter found in the brain and in the sympathetic division of the autonomic nervous system.	4.51 A neurological disease characterized by tremors, rigidity of the limbs, poor balance, and difficulty in initiating movements; caused by degeneration of the nigrostriatal system.
4.61 One of the catecholamines; a hormone secreted by the adrenal medulla; serves also as a neurotransmitter in the brain.	4.52 A drug that blocks the activity of tyrosine hydroxylase and thus interferes with the synthesis of the catecholamines.
4.62 A drug that inhibits the activity of the enzyme dopamine-ß-hydroxylase and thus blocks the production of norepinephrine.	4.53 A drug that interferes with the storage of monoamines in synaptic vesicles.
4.63 A drug that blocks the activity of MAO-A; acts as a noradrenergic agonist.	4.54 A drug that blocks dopamine autoreceptors at low doses; at higher doses blocks postsynaptic receptors as well.
4.64 A dark-colored group of noradrenergic cell bodies located in the pons near the rostral end of the floor of the fourth ventricle.	4.55 A drug that inhibits the reuptake of dopamine.
4.65 Enlarged regions along the length of an axon that contain synaptic vesicles and release a neurotransmitter or neuromodulator.	4.56 A class of enzymes that destroy the monoamines: dopamine, norepinephrine, and serotonin.

4.66 serotonin (5-HT) (*sair a **toe** nin*)	4.75 AMPA receptor
4.67 PCPA	4.76 kainate receptor (***kay** in ate*)
4.68 D system	4.77 metabotropic receptor (*meh tab a **troh** pik*)
4.69 M system	4.78 PCP
4.70 fluoxetine (*floo **ox** i teen*)	4.79 GABA
4.71 fenfluramine (*fen **fluor** i meen*)	4.80 allylglycine
4.72 LSD	4.81 muscimol (***musk** i mawl*)
4.73 glutamate	4.82 bicuculline (*by **kew** kew leen*)
4.74 NMDA receptor	4.83 benzodiazepine (*ben zoe dy **azz** a peen*)

4.75 An ionotropic glutamate receptor that controls a sodium channel; stimulated by AMPA and blocked by CNQX.	4.66 An indolamine transmitter substance; also called 5-hydroxytryptamine.
4.76 An ionotropic glutamate receptor that controls a sodium channel; stimulated by kainic acid and blocked by CNQX.	4.67 A drug that inhibits the activity of tryptophan hydroxylase and thus interferes with the synthesis of 5-HT.
4.77 A metabotropic glutamate receptor.	4.68 A system of serotonergic neurons that originates in the dorsal raphe nucleus; its axonal fibers are thin, with spindle-shaped varicosities that do not appear to form synapses with other neurons.
4.78 Phencyclidine; a drug that binds with the PCP binding site of the NMDA receptor and serves as an inverse agonist.	4.69 A system of serotonergic neurons that originates in the median raphe nucleus; its axonal fibers are thick and rounded and appear to form conventional synapses with other neurons.
4.79 An amino acid; the most important inhibitory transmitter substance in the brain.	4.70 A drug that inhibits the reuptake of 5-HT.
4.80 A drug that inhibits the activity of GAD and thus blocks the synthesis of GABA.	4.71 A drug that stimulates the release of 5-HT.
4.81 A direct agonist for the GABA binding site on the $GABA_A$ receptor.	4.72 A drug that stimulates 5-HT_{2A} receptors.
4.82 A direct antagonist for the GABA binding site on the $GABA_A$ receptor.	4.73 An amino acid; the most important excitatory transmitter substance in the brain.
4.83 A category of anxiolytic drugs; an indirect agonist for the $GABA_A$ receptor.	4.74 A specialized ionotropic glutamate receptor that controls a calcium channel that is normally blocked by Mg^{2+} ions; has several other binding sites.

4.84 anxiolytic (*angz ee oh **lit** ik*)	4.93 caffeine
4.85 ß-CCM	4.94 nitric oxide (NO)
4.86 glycine (***gly** seen*)	4.95 nitric oxide synthase
4.87 strychnine (***strik** neen*)	5.1 experimental ablation
4.88 endogenous opioid (*en **dodge** en us **oh** pee oyd*)	5.2 lesion study
4.89 enkephalin (*en **keff** a lin*)	5.3 excitotoxic lesion (*ek sigh tow **tok** sik*)
4.90 naloxone (*na **lox** own*)	5.4 6-hydroxydopamine (6-HD)
4.91 anandamide (*a **nan** da mide*)	5.5 sham lesion
4.92 adenosine (*a **den** oh seen*)	5.6 stereotaxic surgery (*stair ee oh **tak** sik*)

4.93 A drug that blocks adenosine receptors.	4.84 An anxiety-reducing effect.
4.94 A gas produced by cells in the nervous system; used as a means of communication between cells.	4.85 A direct agonist for the benzodiazepine binding site of the $GABA_A$ receptor.
4.95 The enzyme responsible for the production of nitric oxide.	4.86 An amino acid; an important inhibitory transmitter substance in the lower brain stem and spinal cord.
5.1 The removal or destruction of a portion of the brain of a laboratory animal; presumably, the functions that can no longer be performed are the ones the region previously controlled.	4.87 A direct agonist for the glycine receptor.
5.2 A synonym for experimental ablation.	4.88 A class of peptides secreted by the brain that act as opiates.
5.3 A brain lesion produced by intracerebral injection of an excitatory amino acid, such as kainic acid.	4.89 One of the endogenous opioids.
5.4 A chemical that is selectively taken up by axons and terminal buttons of noradrenergic or dopaminergic neurons and acts as a poison, damaging or killing them.	4.90 A drug that blocks opioid receptors.
5.5 A "placebo" procedure that duplicates all the steps of producing a brain lesion except for the one that actually causes the brain damage.	4.91 A lipid; the endogenous ligand for receptors that bind with THC, the active ingredient of marijuana.
5.6 Brain surgery using a stereotaxic apparatus to position an electrode or cannula in a specified position of the brain.	4.92 A nucleoside; a combination of ribose and adenine; serves as a neuromodulator in the brain.

5.7 bregma	5.16 PHA-L
5.8 stereotaxic atlas	5.17 immunocytochemical method
5.9 stereotaxic apparatus	5.18 retrograde labeling method
5.10 fixative	5.19 fluorogold (*flew roh gold*)
5.11 formalin (*for ma lin*)	5.20 computerized tomography (CT)
5.12 perfusion (*per few zhun*)	5.21 magnetic resonance imaging (MRI)
5.13 microtome (*my krow tome*)	5.22 microelectrode
5.14 scanning electron microscope	5.23 single-unit recording
5.15 anterograde labeling method (*ann ter oh grade*)	5.24 macroelectrode

5.16 Phaseolus vulgaris leukoagglutinin; a protein derived from lima beans used as an anterograde tracer; taken up by dendrites and cell bodies and carried to the ends of the axons.	5.7 The junction of the sagittal and coronal sutures of the skull; often used as a reference point for stereotaxic brain surgery.
5.17 A histological method that uses radioactive antibodies or antibodies bound with a dye molecule to indicate the presence of particular proteins of peptides.	5.8 A collection of drawings of sections of the brain of a particular animal with measurements that provide coordinates for stereotaxic surgery.
5.18 A histological method that labels cell bodies that give rise to the terminal buttons that form synapses with cells in a particular region.	5.9 A device that permits a surgeon to position an electrode or cannula into a specific part of the brain.
5.19 A dye that serves as a retrograde label; taken up by terminal buttons and carried back to the cell bodies.	5.10 A chemical such as formalin; used to prepare and preserve body tissue.
5.20 The use of a device that employs a computer to analyze data obtained by a scanning beam of X rays to produce a two-dimensional picture of a "slice" through the body.	5.11 The aqueous solution of formaldehyde gas; the most commonly used tissue fixative.
5.21 An technique whereby the interior of the body can be accurately imaged; involves the interaction between radio waves and a strong magnetic field.	5.12 The process by which an animal's blood is replaced by a fluid such as a saline solution or a fixative in preparing the brain for histological examination.
5.22 A very fine electrode, generally used to record activity of individual neurons.	5.13 An instrument that produces very thin slices of body tissues.
5.23 Recording of the electrical activity of a single neuron.	5.14 A microscope that provides three-dimensional information about the shape of the surface of a small object.
5.24 An electrode used to record the electrical activity of large numbers of neurons in a particular region of the brain; much larger than a microelectrode.	5.15 A histological method that labels the axons and terminal buttons of neurons whose cell bodies are located in a particular region.

5.25 electroencephalogram (EEG)	5.34 in situ hybridization (*in see too*)
5.26 2-deoxyglucose (2-DG) (*dee ox ee **gloo** kohss*)	5.35 double labeling
5.27 autoradiography	6.1 sensory receptor
5.28 Fos (*fahs*)	6.2 sensory transduction
5.29 positron emission tomography (PET)	6.3 receptor potential
5.30 functional MRI (fMRI)	6.4 hue
5.31 microdialysis	6.5 brightness
5.32 multi-barreled micropipette	6.6 saturation
5.33 microiontophoresis	6.7 vergence movement

5.34 The production of DNA complementary to a particular messenger RNA in order to detect the presence of the RNA.	**5.25** An electrical brain potential recorded by placing electrodes on in the scalp.
5.35 Labeling neurons in a particular region by two different means; for example, by using an anterograde tracer and a label for a particular enzyme.	**5.26** A sugar that enters cells along with glucose but is not metabolized.
6.1 A specialized neuron that detects a particular category of physical events.	**5.27** A procedure that locates radioactive substances in a slice of tissue; the radiation exposes a photographic emulsion or a piece of film that covers the tissue.
6.2 The process by which sensory stimuli are transduced into slow, graded receptor potentials.	**5.28** A protein produced in the nucleus of a neuron in response to synaptic stimulation.
6.3 A slow, graded electrical potential produced by a receptor cell in response to a physical stimulus.	**5.29** The use of a device that reveals the localization of a radioactive tracer in a living brain.
6.4 One of the perceptual dimensions of color; the dominant wavelength.	**5.30** A modification of the MRI procedure that permits the measurement of regional metabolism in the brain.
6.5 One of the perceptual dimensions of color; intensity.	**5.31** A procedure for analyzing chemicals present in the interstitial fluid through a small piece of tubing made of a semipermeable membrane that is implanted in the brain.
6.6 One of the perceptual dimensions of color; purity.	**5.32** A group of micropipettes attached together, used to infuse several different substances by means of iontophoresis while recording from a single neuron.
6.7 The cooperative movement of the eyes, which ensures that the image of an object falls on identical portions of both retinas.	**5.33** A procedure that uses electricity to eject a chemical from a micropipette in order to determine the effects of the chemical on the electrical activity of a cell.

6.8 saccadic movement (*suh **kad** ik*)	6.17 bipolar cell
6.9 pursuit movement	6.18 ganglion cell
6.10 accommodation	6.19 horizontal cell
6.11 retina	6.20 amacrine cell (***amm** a krin*)
6.12 rod	6.21 lamella
6.13 cone	6.22 photopigment
6.14 photoreceptor	6.23 opsin (***opp** sin*)
6.15 fovea (***foe** vee a*)	6.24 retinal (***rett** i nahl*)
6.16 optic disk	6.25 rhodopsin (*roh **dopp** sin*)

6.17 A bipolar neuron located in the middle layer of the retina, conveying information from the photoreceptors to the ganglion cells.	6.8 The rapid, jerky movement of the eyes used in scanning a visual scene.
6.18 A neuron located in the retina that receives visual information from bipolar cells; its axons give rise to the optic nerve.	6.9 The movement that the eyes make to maintain an image on the fovea.
6.19 A neuron in the retina that interconnects adjacent photoreceptors and the outer processes of the bipolar cells.	6.10 Changes in the thickness of the lens of the eye, accomplished by the ciliary muscles, that focus images of near or distant objects on the retina.
6.20 A neuron in the retina that interconnects adjacent ganglion cells and the inner processes of the bipolar cells.	6.11 The neural tissue and photoreceptive cells located on the inner surface of the posterior portion of the eye.
6.21 A layer of membrane containing photopigments; found in rods and cones of the retina.	6.12 One of the receptor cells of the retina; sensitive to light of low intensity.
6.22 A protein dye bonded to retinal, a substance derived from vitamin A; responsible for transduction of visual information.	6.13 One of the receptor cells of the retina; maximally sensitive to one of three different wavelengths of light and hence encodes color vision.
6.23 A class of protein that, together with retinal, constitutes the photopigments.	6.14 One of the receptor cells of the retina; transduces photic energy into electrical potentials.
6.24 A chemical synthesized from vitamin A; joins with an opsin to form a photopigment.	6.15 The region of the retina that mediates the most acute vision of birds and higher mammals. Color-sensitive cones constitute the only type of photoreceptor found in the fovea.
6.25 A particular opsin found in rods.	6.16 The location of the exit point from the retina of the fibers of the ganglion cells that form the optic nerve; responsible for the blind spot.

6.26 transducin	6.35 deuteranopia (*dew ter an* **owe** *pee a*)
6.27 dorsal lateral geniculate nucleus	6.36 tritanopia (*try tan* **owe** *pee a*)
6.28 magnocellular layer	6.37 negative afterimage
6.29 parvocellular layer	6.38 complementary colors
6.30 calcarine fissure (**kal** *ka rine*)	6.39 simple cell
6.31 striate cortex (**stry** *ate*)	6.40 complex cell
6.32 optic chiasm	6.41 sine-wave grating
6.33 receptive field	6.42 spatial frequency
6.34 protanopia (*pro tan* **owe** *pee a*)	6.43 retinal disparity

6.35 An inherited form of defective color vision in which red and green hues are confused; "green" cones are filled with "red" cone opsin.	**6.26** A G protein that is activated when a photon strikes a photopigment; activates phosphodiesterase molecules, which destroy cyclic GMP and close cation channels in the photoreceptor.
6.36 An inherited form of defective color vision in which hues with short wavelengths are confused; "blue" cones are either lacking or faulty.	**6.27** A group of cell bodies within the lateral geniculate body of the thalamus; receives inputs from the retina and projects to the primary visual cortex.
6.37 The image seen after a portion of the retina is exposed to an intense visual stimulus; consists of colors complementary to those of the physical stimulus.	**6.28** One of the inner two layers of cells in the dorsal lateral geniculate nucleus; transmits information necessary for the perception of form, movement, depth, and small differences in brightness.
6.38 Colors that make white or gray when mixed together.	**6.29** One of the four outer layers of cells in the dorsal lateral geniculate nucleus; transmits information necessary for perception of color and fine details.
6.39 An orientation-sensitive neuron in the striate cortex whose receptive field is organized in an opponent fashion.	**6.30** A horizontal fissure on the inner surface of the posterior cerebral cortex; the location of the primary visual cortex.
6.40 A neuron in the visual cortex that responds to the presence of a line segment with a particular orientation located within its receptive field, especially when the line moves perpendicularly to its orientation.	**6.31** The primary visual cortex.
6.41 A series of straight parallel bands varying continuously in brightness according to a sine-wave function, along a line perpendicular to their lengths.	**6.32** A cross-shaped connection between the optic nerves, located below the base of the brain, just anterior to the pituitary gland.
6.42 The relative width of the bands in a sine-wave grating, measured in cycles per degree of visual angle.	**6.33** That portion of the visual field in which the presentation of visual stimuli will produce an alteration in the firing rate of a particular neuron.
6.43 The fact that points on objects located at different distances from the observer will fall on slightly different locations on the two retinas; provides the basis for stereopsis.	**6.34** An inherited form of defective color vision in which red and green hues are confused; "red" cones are filled with "green" cone opsin.

6.44 cytochrome oxidase (CO) blob	6.53 prosopagnosia (*prah soh pag **no** zha*)
6.45 ocular dominance	6.54 associative visual agnosia
6.46 blindsight	6.55 pulvinar (***pull** vi nar*)
6.47 extrastriate cortex	6.56 Balint's syndrome
6.48 color constancy	6.57 optic ataxia (*ay **tack** see a*)
6.49 achromatopsia (*ay krohm a **top** see a*)	6.58 ocular apraxia (*ay **prak** see a*)
6.50 inferior temporal cortex	6.59 simultanagnosia (*sime ul tane ag **no** zha*)
6.51 visual agnosia (*ag **no** zha*)	7.1 pitch
6.52 apperceptive visual agnosia	7.2 hertz (Hz)

6.53 Failure to recognize particular people by the sight of their faces.	6.44 The central region of a module of the primary visual cortex, revealed by a stain for cytochrome oxidase; contains wavelength-sensitive neurons; part of the parvocellular system.
6.54 Inability to identify objects that are perceived visually, even though the form of the perceived object can be drawn or matched with similar objects.	6.45 The extent to which a particular neuron receives more input from one eye than from the other.
6.55 A large thalamic nucleus that projects to the visual association cortex and may play a role in compensating for eye and head movements.	6.46 The ability of a person to reach for objects located in his or her "blind" field; occurs after damage restricted to the primary visual cortex.
6.56 A syndrome caused by bilateral damage to the parieto-occipital region; includes optic ataxia, ocular apraxia, and simultanagnosia	6.47 A region of visual association cortex; receives fibers from the striate cortex and from the superior colliculi and projects to the inferior temporal cortex.
6.57 Difficulty in reaching for objects under visual guidance.	6.48 The relatively constant appearance of the colors of objects viewed under varying lighting conditions.
6.58 Difficulty in visual scanning.	6.49 Inability to discriminate among different hues; caused by damage to the visual association cortex.
6.59 Difficulty in perceiving more than one object at a time.	6.50 In primates, the highest level of visual association cortex; located on the inferior portion of the temporal lobe.
7.1 A perceptual dimension of sound; corresponds to the fundamental frequency.	6.51 Deficits in visual perception in the absence of blindness; caused by brain damage.
7.2 Cycles per second.	6.52 Failure to perceive objects, even though visual acuity is relatively normal.

7.3		7.12	
loudness		organ of Corti	

7.4		7.13	
timbre (*tim ber* or *tamm ber*)		hair cell	

7.5		7.14	
tympanic membrane		Deiters's cell (*dye terz*)	

7.6		7.15	
ossicle (*ahss i kul*)		basilar membrane (*bazz i ler*)	

7.7		7.16	
malleus		tectorial membrane (*tek torr ee ul*)	

7.8		7.17	
incus		round window	

7.9		7.18	
stapes (*stay peez*)		cilium	

7.10		7.19	
cochlea (*cock lee uh*)		tip link	

7.11		7.20	
oval window		insertional plaque	

7.12 The sensory organ on the basilar membrane that contains the auditory hair cells.	**7.3** A perceptual dimension of sound; corresponds to intensity.
7.13 The receptive cell of the auditory apparatus.	**7.4** A perceptual dimension of sound; corresponds to complexity.
7.14 A supporting cell found in the organ of Corti; sustains the auditory hair cells.	**7.5** The eardrum.
7.15 A membrane in the cochlea of the inner ear; contains the organ of Corti.	**7.6** One of the three bones of the middle ear.
7.16 A membrane located above the basilar membrane; serves as a shelf against which the cilia of the auditory hair cells move.	**7.7** The "hammer"; the first of the three ossicles.
7.17 An opening in the bone surrounding the cochlea of the inner ear that permits vibrations to be transmitted, via the oval window, into the fluid in the cochlea.	**7.8** The "anvil"; the second of the three ossicles.
7.18 A hairlike appendage of a cell involved in movement or in transducing sensory information; found on the receptors in the auditory and vestibular system.	**7.9** The "stirrup" the last of the three ossicles.
7.19 Elastic filaments that attach the tip of one cilium to the side of the adjacent cilium.	**7.10** The snail-shaped structure of the inner ear that contains the auditory transducing mechanisms.
7.20 The point of attachment of a tip link to a cilium.	**7.11** An opening in the bone surrounding the cochlea that reveals a membrane, against which the baseplate of the stapes presses, transmitting sound vibrations into the fluid within the cochlea.

7.21 cochlear nerve	7.30 fundamental frequency
7.22 olivocochlear bundle	7.31 overtone
7.23 cochlear nucleus	7.32 phase difference
7.24 superior olivary complex	7.33 vestibular sac
7.25 lateral lemniscus	7.34 semicircular canal
7.26 tonotopic representation (*tonn oh **top** ik*)	7.35 utricle (***you** trih kul*)
7.27 place code	7.36 saccule (***sak** yule*)
7.28 cochlear implant	7.37 ampulla (*am **pull** uh*)
7.29 rate code	7.38 cupula (***kew** pew luh*)

7.30 The lowest, and usually most intense, frequency of a complex sound; most often perceived as the sound's basic pitch.	7.21 The branch of the auditory nerve that transmits auditory information from the cochlea to the brain.
7.31 The frequency of complex tones that occurs at multiples of the fundamental frequency.	7.22 A bundle of efferent axons that travel from the olivary complex of the medulla to the auditory hair cells on the cochlea.
7.32 The difference in arrival times of sound waves at each of the eardrums.	7.23 One of a group of nuclei in the medulla that receive auditory information from the cochlea.
7.33 One of a set of two receptor organs in each inner ear that detect changes in the tilt of the head.	7.24 A group of nuclei in the medulla; involved with auditory functions, including localization of the source of sounds.
7.34 One of the three ringlike structures of the vestibular apparatus that detect changes in head rotation.	7.25 A band of fibers running rostrally through the medulla and pons; carries fibers of the auditory system.
7.35 One of the vestibular sacs.	7.26 A topographically organized mapping of different frequencies of sound that are represented in a particular region of the brain.
7.36 One of the vestibular sacs.	7.27 The system by which information about different frequencies is coded by different locations on the basilar membrane.
7.37 An enlargement in a semicircular canal; contains the cupula and the crista.	7.28 An electronic device surgically implanted in the inner ear that can enable deaf people to hear.
7.38 A gelatinous mass found in the ampulla of the semicircular canals; moves in response to the flow of the fluid in the canals.	7.29 The system by which information about different frequencies is coded by the rate of firing of neurons in the auditory system.

7.48 A member of a family of fatty acid derivatives that serve as hormones; first discovered in the prostate gland; involved in many physiological processes, including pain perception.	**7.39** A nodule on the vestibular nerve that contains the cell bodies of the bipolar neurons that convey vestibular information to the brain.
7.49 Sensations that appear to originate in a limb that has been amputated.	**7.40** One of the somatosenses; includes sensitivity to stimuli that involve the skin.
7.50 A nucleus of the raphe that contains serotonin-secreting neurons that project to the dorsal gray matter of the spinal cord via the dorsolateral columns and is involved in analgesia produced by opiates.	**7.41** Perception of the body's own movements.
7.51 A G protein that plays a vital role in the transduction of sweetness and bitterness.	**7.42** A sense modality that arises from receptors located within the inner organs of the body.
7.52 The taste sensation produced by glutamate.	**7.43** Skin that does not contain hair; found on the palms and soles of the feet.
7.53 A branch of the facial nerve that passes beneath the eardrum; conveys taste information from the anterior part of the tongue and controls the secretion of some salivary glands.	**7.44** A vibration-sensitive organ located in hairy skin.
7.54 A nucleus of the medulla that receives information from visceral organs and from the gustatory system.	**7.45** A specialized, encapsulated somatosensory nerve ending that detects mechanical stimuli, especially vibrations.
7.55 The epithelial tissue of the nasal sinus that covers the cribriform plate; contains the cilia of the olfactory receptors.	**7.46** The touch-sensitive end organs located in the papillae, small elevations of the dermis that project up into the epidermis.
7.56 The protrusion at the end of the olfactory nerve; receives input from the olfactory receptors.	**7.47** The touch-sensitive end organs found at the base of the epidermis, adjacent to sweat ducts.

7.39 vestibular ganglion	7.48 prostaglandin
7.40 cutaneous sense (*kew **tane** ee us*)	7.49 phantom limb
7.41 kinesthesia	7.50 nucleus raphe magnus
7.42 organic sense	7.51 gustducin (*gust **doo** sin*)
7.43 glabrous skin (***glab** russ*)	7.52 umami (*oo mah mee*)
7.44 Ruffini corpuscle	7.53 chorda tympani
7.45 Pacinian corpuscle (*pa **chin** ee un*)	7.54 nucleus of the solitary tract (NST)
7.46 Meissner's corpuscle	7.55 olfactory epithelium
7.47 Merkel's disk	7.56 olfactory bulb

7.57 mitral cell	8.8 motor unit
7.58 olfactory glomerulus (*glow **mare** you luss*)	8.9 myofibril
8.1 skeletal muscle	8.10 actin
8.2 flexion	8.11 myosin
8.3 extension	8.12 striated muscle
8.4 extrafusal muscle fiber	8.13 neuromuscular junction
8.5 alpha motor neuron	8.14 motor endplate
8.6 intrafusal muscle fiber	8.15 endplate potential
8.7 gamma motor neuron	8.16 Golgi tendon organ (GTO)

8.8 A motor neuron and its associated muscle fibers.	**7.57** A neuron located in the olfactory bulb that receives information from olfactory receptors; axons of mitral cells bring information to the rest of the brain.
8.9 An element of muscle fibers that consists of overlapping strands of actin and myosin; responsible for muscular contractions.	**7.58** A bundle of dendrites of mitrial cells and the associated terminal buttons of the axons of olfactory receptors.
8.10 One of the proteins (with myosin) that provide the physical basis for muscular contraction.	**8.1** One of the striated muscles attached to bones.
8.11 One of the proteins (with actin) that provide the physical basis for muscular contraction.	**8.2** A movement of a limb that tends to bend its joints; opposite of extension.
8.12 Skeletal muscle; muscle that contains striations.	**8.3** A movement of a limb that tends to straighten its joints; the opposite of flexion.
8.13 The synapse between the terminal buttons of an axon and a muscle fiber.	**8.4** One of the muscle fibers that are responsible for the force exerted by contraction of a skeletal muscle.
8.14 The postsynaptic membrane of a neuromuscular junction.	**8.5** A neuron whose axon forms synapses with extrafusal muscle fibers of a skeletal muscle; activation contracts the muscle fibers.
8.15 The postsynaptic potential that occurs in the motor endplate in response to release of acetylcholine by the terminal button.	**8.6** A muscle fiber that functions as a stretch receptor, arranged parallel to the extrafusal muscle fibers, thus detecting changes in muscle length.
8.16 The receptor organ at the junction of the tendon and muscle that is sensitive to stretch.	**8.7** A neuron whose axons form synapses with intrafusal muscle fibers.

8.17 smooth muscle	8.26 supplementary motor area
8.18 cardiac muscle	8.27 premotor cortex
8.19 monosynaptic stretch reflex	8.28 prefrontal cortex
8.20 decerebrate	8.29 lateral group
8.21 decerebrate rigidity	8.30 ventromedial group
8.22 clasp-knife reflex	8.31 corticospinal tract
8.23 agonist	8.32 pyramidal tract
8.24 antagonist	8.33 lateral corticospinal tract
8.25 somatotopic organization	8.34 ventral corticospinal tract

8.26 A region of motor association cortex of the dorsal and dorsomedial frontal lobe, rostral to the primary motor cortex.	8.17 Nonstriated muscle innervated by the autonomic nervous system, found in the walls of blood vessels, in the reproductive tracts, in sphincters, within the eye, in the digestive system, and around hair follicles.
8.27 A region of motor association cortex of the lateral frontal lobe, rostral to the primary motor cortex.	8.18 The muscle responsible for the contraction of the heart.
8.28 The neocortex of the frontal lobes rostral to the supplementary motor area and premotor cortex.	8.19 A reflex in which a muscle contracts in response to its being quickly stretched; involves a sensory neuron and a motor neuron, with one synapse between them.
8.29 The corticospinal tract, the corticobulbar tract, and the rubrospinal tract.	8.20 Describes an animal whose brain stem has been transected.
8.30 The vestibulospinal tract, the tectospinal tract, the reticulospinal tract, and the ventral corticospinal tract.	8.21 Simultaneous contraction of agonistic and antagonistic muscles; caused by decerebration or damage to the reticular formation.
8.31 The system of axons that originates in the motor cortex and terminates in the ventral gray matter of the spinal cord.	8.22 A reflex that occurs when force is applied to flex or extend the limb of an animal showing decerebrate rigidity; resistance is replaced by sudden relaxation.
8.32 An alternate term for the corticospinal tract.	8.23 A muscle whose contraction produces or facilitates a particular movement.
8.33 The system of axons that originates in the motor cortex and terminates in the contralateral ventral gray matter of the spinal cord; controls movements of the distal limbs.	8.24 A muscle whose contraction resists or reverses a particular movement.
8.34 The system of axons that originates in the motor cortex and terminates in the ipsilateral ventral gray matter of the spinal cord; controls movements of the upper legs and trunk.	8.25 A topographically organized mapping of parts of the body that are represented in a particular region of the brain.

8.35 corticobulbar pathway	8.44 left parietal apraxia
8.36 rubrospinal tract	8.45 constructional apraxia
8.37 corticorubral tract	8.46 caudate nucleus
8.38 vestibulospinal tract	8.47 putamen
8.39 tectospinal tract	8.48 globus pallidus
8.40 reticulospinal tract	8.49 ventral anterior nucleus (of thalamus)
8.41 apraxia	8.50 ventrolateral nucleus (of thalamus)
8.42 callosal apraxia	8.51 Huntington's chorea
8.43 sympathetic apraxia	8.52 flocculonodular lobe

8.44 An apraxia caused by damage to the left parietal lobe; characterized by difficulty in producing sequences of movements by verbal request or in imitation of movements made by someone else.	8.35 A bundle of axons from the motor cortex to the fifth, seventh, ninth, tenth, eleventh, and twelfth cranial nerves; controls movements of the face, neck, tongue, and parts of the extraocular eye muscles.
8.45 Difficulty in drawing pictures or diagrams or in making geometrical constructions of elements such as building blocks or sticks; caused by damage to the right parietal lobe.	8.36 The system of axons that travels from the red nucleus to the spinal cord; controls independent limb movements.
8.46 A telencephalic nucleus; one of the input nuclei of basal ganglia along with the putamen; involved with control of voluntary movement.	8.37 The system of axons that travels from the motor cortex to the red nucleus.
8.47 A telencephalic nucleus; one of the input nuclei of the basal ganglia along with the caudate nucleus; involved with control of voluntary movement.	8.38 A bundle of axons that travels from the vestibular nuclei to the gray matter of the spinal cord; controls postural movements in response to information from the vestibular system.
8.48 A telencephalic nucleus; the primary output nucleus of the basal ganglia; involved with control of voluntary movement.	8.39 A bundle of axons that travels from the tectum to the spinal cord; coordinates head and trunk movements with eye movements.
8.49 One of the two thalamic nuclei that receive projections from the basal ganglia and send projections to the motor cortex.	8.40 A bundle of axons that travels from the reticular formation to the gray matter of the spinal cord; controls the muscles responsible for postural movements.
8.50 One of the two thalamic nuclei that receive projections from the basal ganglia and send projections to the motor cortex.	8.41 Difficulty in carrying out purposeful movements, in the absence of paralysis or muscular weakness.
8.51 An fatal inherited disorder that causes degeneration of the caudate nucleus and putamen; characterized by uncontrollable jerking movements, writhing movements, and dementia.	8.42 An apraxia of the left hand caused by damage to the anterior corpus callosum.
8.52 A region of the cerebellum; involved in control of postural reflexes.	8.43 A movement disorder of the left hand caused by damage to the left frontal lobe; similar to callosal apraxia.

8.53 vermis	9.4 beta activity
8.54 fastigial nucleus	9.5 synchrony
8.55 interposed nuclei	9.6 desynchrony
8.56 pontine nucleus	9.7 theta activity
8.57 dentate nucleus	9.8 delta activity
8.58 mesencephalic locomotor region	9.9 REM sleep
9.1 electromyogram (EMG) (*my oh gram*)	9.10 non-REM sleep
9.2 electro-oculogram (EOG) (*ah kew loh gram*)	9.11 slow-wave sleep
9.3 alpha activity	9.12 basic rest-activity cycle (BRAC)

9.4 Irregular electrical activity of 13-30 Hz recorded from the brain; generally associated with a state of arousal.	8.53 The portion of the cerebellum located at the midline; receives somatosensory information and helps control the vestibulospinal and reticulospinal tracts through its connections with the fastigial nucleus.
9.5 High-voltage, low-frequency EEG activity, characteristic of slow-wave sleep or coma, during which neurons fire together in a regular fashion.	8.54 A deep cerebellar nucleus; involved in the control of movement by the reticulospinal and vestibulospinal tracts.
9.6 Irregular electrical activity recorded from the brain, generally associated with periods of arousal.	8.55 A set of deep cerebellar nuclei; involved in the control of the rubrospinal system.
9.7 EEG activity of 5-8 Hz that occurs intermittently during early stages of slow-wave sleep and REM sleep.	8.56 A large nucleus in the pons that serves as an important source of input to the cerebellum.
9.8 Regular, synchronous electrical activity of approximately 1-4 Hz recorded from the brain; occurs during the deepest stages of slow-wave sleep.	8.57 A deep cerebellar nucleus; involved in the control of rapid, skilled movements by the corticospinal and rubrospinal systems.
9.9 A period of desynchronized EEG activity during sleep, at which time dreaming, rapid eye movements, and muscular paralysis occur; also called *paradoxical sleep.*	8.58 A region of the reticular formation of the midbrain whose stimulation causes alternating movements of the limbs normally seen during locomotion.
9.10 All stages of sleep except REM sleep.	9.1 An electrical potential recorded from an electrode placed on or in a muscle.
9.11 Non-REM sleep, characterized by synchronized EEG activity during its deeper stages.	9.2 An electrical potential from the eyes, recorded by means of electrodes placed on the skin around them; detects eye movements.
9.12 A 90-min cycle (in humans) of waxing and waning alertness, controlled by a biological clock in the caudal brain stem; controls cycles of REM sleep and slow-wave sleep.	9.3 Smooth electrical activity of 8-12 Hz recorded from the brain; generally associated with a state of relaxation.

9.13 fatal familial insomnia	9.22 medial pontine reticular formation (MPRF)
9.14 rebound phenomenon	9.23 magnocellular nucleus
9.15 locus coeruleus (*sa **roo** lee us*)	9.24 drug dependency insomnia
9.16 raphe nuclei (*ruh **fay***)	9.25 sleep apnea (***app** nee a*)
9.17 basal forebrain region	9.26 narcolepsy (***nahr** ko lep see*)
9.18 POAH	9.27 sleep attack
9.19 PGO wave	9.28 cataplexy (***kat** a plex ee*)
9.20 peribrachial area (*pair ee **bray** kee ul*)	9.29 sleep paralysis
9.21 carbachol (***car** ba call*)	9.30 hypnagogic hallucination (*hip na **gah** jik*)

9.22 A region that contains neurons involved in the initiation of REM sleep; activated by acetylcholinergic neurons of the peribrachial area.	9.13 A fatal inherited disorder characterized by progressive insomnia.
9.23 A nucleus in the medulla; involved in the atonia (muscular paralysis) that accompanies REM sleep.	9.14 The increased frequency or intensity of a phenomenon after it has been temporarily suppressed; for example, the increase in REM sleep seen after a period of REM sleep deprivation.
9.24 An insomnia caused by the side effects of ever-increasing doses of sleeping medications.	9.15 A dark-colored group of noradrenergic cell bodies located in the pons near the rostral end of the floor of the fourth ventricle; involved in arousal and vigilance.
9.25 Cessation of breathing while sleeping.	9.16 A group of nuclei located in the reticular formation of the medulla, pons, and midbrain, situated along the midline; contain serotonergic neurons.
9.26 A sleep disorder characterized by periods of irresistible sleep, attacks of cataplexy, sleep paralysis, and hypnagogic hallucinations.	9.17 The region at the base of the forebrain rostral to the hypothalamus; involved in thermoregulation and control of sleep.
9.27 A symptom of narcolepsy; an irresistible urge to sleep during the day, after which the person awakes feeling refreshed.	9.18 The region of the preoptic area and the adjacent anterior hypothalamus, involved in thermoregulation and induction of slow-wave sleep.
9.28 A symptom of narcolepsy; complete paralysis that occurs during waking.	9.19 Bursts of phasic electrical activity originating in the pons, followed by activity in the lateral geniculate nucleus and visual cortex; a characteristic of REM sleep.
9.29 A symptom of narcolepsy; paralysis occurring just before a person falls asleep.	9.20 The region around the brachium conjunctivum, located in the dorsolateral pons; contains acetylcholinergic neurons involved in the initiation of REM sleep.
9.30 A symptom of narcolepsy; vivid dreams that occur just before a person falls asleep; accompanied by sleep paralysis.	9.21 A drug that stimulates acetylcholine receptors.

9.31 REM without atonia (*ay **tone** ee a*)	10.2 gamete (***gamm** eet*)
9.32 circadian rhythm (*sur **kay** dee un* or *sur ka **dee** un*)	10.3 sex chromosome
9.33 zeitgeber (***tsite** gay ber*)	10.4 gonad (rhymes with ***moan** ad*)
9.34 suprachiasmatic nucleus (SCN) (*soo pra ky az **mat** ik*)	10.5 organizational effect (of hormone)
9.35 intergeniculate leaflet (IGL)	10.6 activational effect (of hormone)
9.36 neuropeptide Y	10.7 Müllerian system
9.37 pineal gland (*py **nee** ul*)	10.8 Wolffian system
9.38 melatonin (*mell a **tone** in*)	10.9 anti-Müllerian hormone
10.1 sexually dimorphic behavior	10.10 defeminizing effect

10.2 A mature reproductive cell; a sperm or ovum.	9.31 A neurological disorder in which the person does not become paralyzed during REM sleep and thus acts out dreams.
10.3 The X and Y chromosomes, which determine an organism's gender. Normally, XX individuals are female, and XY individuals are male.	9.32 A daily rhythmical change in behavior or physiological process.
10.4 An ovary or testis.	9.33 A stimulus (usually the light of dawn) that resets the biological clock responsible for circadian rhythms.
10.5 The effect of a hormone on tissue differentiation and development.	9.34 A nucleus situated atop the optic chiasm. It contains a biological clock responsible for organizing many of the body's circadian rhythms.
10.6 The effect of a hormone that occurs in the fully developed organism; may depend on the organism's prior exposure to the organizational effects of hormones.	9.35 A part of the lateral geniculate nucleus that receives information from the retina and projects to the SCN.
10.7 The embryonic precursors of the female internal sex organs.	9.36 A peptide released by the terminals of the neurons that project from the IGL to the SCN.
10.8 The embryonic precursors of the male internal sex organs.	9.37 A gland attached to the dorsal tectum; produces melatonin and plays a role in circadian and seasonal rhythms.
10.9 A peptide secreted by the fetal testes that inhibits the development of the Müllerian system, which would otherwise become the female internal sex organs.	9.38 A hormone secreted during the night by the pineal body; plays a role in circadian and seasonal rhythms.
10.10 An effect of a hormone present early in development that reduces or prevents the later development of anatomical or behavioral characteristics typical of females.	10.1 A behavior that has different forms or that occurs with different probabilities or under different circumstances in males and females.

10.11 androgen (*an dro jen*)	10.20 follicle-stimulating hormone (FSH)
10.12 masculinizing effect	10.21 luteinizing hormone (LH) (*lew tee a nize ing*)
10.13 testosterone (*tess **tahss** ter own*)	10.22 estradiol (*ess tra **dye** ahl*)
10.14 dihydrotestosterone (*dy hy dro tess **tahss** ter own*)	10.23 estrogen (***ess** trow jen*)
10.15 androgen insensitivity syndrome	10.24 menstrual cycle (***men** strew al*)
10.16 persistent Müllerian duct syndrome	10.25 estrous cycle
10.17 Turner's syndrome	10.26 ovarian follicle
10.18 gonadotropin releasing hormone (*go **nad** oh trow pin*)	10.27 corpus luteum (***lew** tee um*)
10.19 gonadotropic hormone	10.28 progesterone (*pro **jess** ter own*)

10.20 The hormone of the anterior pituitary gland that causes development of an ovarian follicle and the maturation of its oocyte into an ovum.	10.11 A male sex steroid hormone. Testosterone is the principal mammalian androgen.
10.21 A hormone of the anterior pituitary gland that causes ovulation and development of the ovarian follicle into a corpus luteum.	10.12 An effect of a hormone present early in development that promotes the later development of anatomical or behavioral characteristics typical of males.
10.22 The principal estrogen of many mammals, including humans.	10.13 The principal androgen found in males.
10.23 A class of sex hormones that cause maturation of the female genitalia, growth of breast tissue, and development of other physical features characteristic of females.	10.14 An androgen, produced from testosterone through the action of the enzyme 5Ó reductase.
10.24 The female reproductive cycle of most primates, including humans; characterized by growth of the lining of the uterus, ovulation, development of a corpus luteum, and (if pregnancy does not occur), menstruation.	10.15 A condition caused by a congenital lack of functioning androgen receptors; in a person with XY sex chromosomes, causes the development of a female with testes but no internal sex organs.
10.25 The female reproductive cycle of mammals other than primates.	10.16 A condition caused by a congenital lack of functioning anti-Müllerian hormone receptors; in a male, causes development of both male and female internal sex organs.
10.26 A cluster of epithelial cells surrounding an oocyte, which develops into an ovum.	10.17 The presence of only one sex chromosome (an X chromosome); characterized by lack of ovaries but otherwise normal female sex organs and genitalia.
10.27 A cluster of cells that develops from the ovarian follicle after ovulation; secretes estradiol and progesterone.	10.18 A hypothalamic hormone that stimulates the anterior pituitary gland to secrete gonadotropic hormone.
10.28 A steroid hormone produced by the ovary that maintains the endometrial lining of the uterus during the later part of the menstrual cycle and during pregnancy; along with estradiol, it promotes receptivity in female mammals with estrous cycles.	10.19 A hormone of the anterior pituitary gland that has a stimulating effect on cells of the gonads.

10.29 refractory period (*ree **frak** to ree*)	10.38 Vandenbergh effect
10.30 Coolidge effect	10.39 Bruce effect
10.31 aromatization (***air** oh mat i **zay** shun*)	10.40 vomeronasal organ (*voah mer oh **nay** zul*)
10.32 oxytocin (*ox ee **tow** sin*)	10.41 accessory olfactory bulb
10.33 prolactin	10.42 medial nucleus of the amygdala (*a **mig** da la*)
10.34 lordosis	10.43 congenital adrenal hyperplasia (CAH) (*hy per **play** zha*)
10.35 pheromone (***fair** oh moan*)	10.44 spinal nucleus of the bulbocavernosus (SNB) (*bul bo kav er **no** sis*)
10.36 Lee-Boot effect	10.45 medial preoptic area (MPA)
10.37 Whitten effect	10.46 sexually dimorphic nucleus

10.38 The earlier onset of puberty seen in female animals that are housed with males; caused by a pheromone in the male's urine; first observed in mice.	**10.29** A period of time after a particular action (for example, an ejaculation by a male) during which that action cannot occur again.
10.39 Termination of pregnancy caused by the odor of a pheromone in the urine of a male other than the one that impregnated the female; first identified in mice.	**10.30** The restorative effect of introducing a new female sex partner to a male that has apparently become "exhausted" by sexual activity.
10.40 A sensory organ that detects the presence of certain chemicals, especially when a liquid is actively sniffed; mediates the effects of some pheromones.	**10.31** A chemical reaction catalyzed by an aromatase; the process by which testosterone is transformed into estradiol.
10.41 A neural structure located in the main olfactory bulb that receives information from the vomeronasal organ.	**10.32** A hormone secreted by the posterior pituitary gland; causes contraction of the smooth muscle of the milk ducts, the uterus, and the male ejaculatory system; also serves as a neurotransmitter in the brain.
10.42 A nucleus that receives olfactory information from the olfactory bulb and accessory olfactory bulb; involved in the effects of odors and pheromones on reproductive behavior.	**10.33** A hormone of the anterior pituitary gland, necessary for production of milk; has an inhibitory effect on male sexual behavior.
10.43 A condition characterized by hypersecretion of androgens by the adrenal cortex; in females, causes masculinization of the external genitalia.	**10.34** A spinal sexual reflex seen in many four-legged female mammals; arching of the back in response to approach of a male or to touching the flanks, which elevates the hindquarters.
10.44 A nucleus located in the lower spinal cord; in some species of rodents, present only in males.	**10.35** A chemical released by one animal that affects the behavior or physiology of another animal; usually smelled or tasted.
10.45 An area of cell bodies just rostral to the hypothalamus; plays an essential role in male sexual behavior.	**10.36** The increased incidence of false pregnancies seen in female animals that are housed together; caused by a pheromone in the animals' urine; first observed in mice.
10.46 A nucleus in the preoptic area that is much larger in males than in females; first observed in rats; plays a role in male sexual behavior.	**10.37** The synchronization of the menstrual or estrous cycles of a group of females, which occurs only in the presence of a pheromone in a male's urine.

10.47 vasopressin (*vay zo* **press** *in*)	11.4 basal nucleus
10.48 ventromedial nucleus of the hypothalamus (VMH)	11.5 conditioned emotional response
10.49 periaqueductal gray matter (PAG)	11.6 coping response
10.50 parturition (*par tew* **ri** *shun*)	11.7 orbitofrontal cortex
10.51 stria terminalis (**stree** *a ter mi* **nal** *is*)	11.8 akinetic mutism
10.52 ventral tegmental area	11.9 display rule
11.1 medial nucleus	11.10 volitional facial paresis
11.2 lateral/basolateral nuclei	11.11 emotional facial paresis
11.3 central nucleus	11.12 Wada test

11.4 A group of subnuclei of the amygdala that receives sensory input from the lateral and basolateral nuclei and relays information to other amygdaloid nuclei and to the periaqueductal gray matter.	10.47 A hormone secreted by the posterior pituitary gland that controls the secretion of urine by the kidneys; also serves as a neurotransmitter in the brain.
11.5 A classically conditioned response that occurs when a neutral stimulus is followed by an aversive stimulus; usually includes autonomic, behavioral, and endocrine components such as changes in heart rate, freezing, and secretion of stress-related hormones.	10.48 A large nucleus of the hypothalamus located near the walls of the third ventricle; plays an essential role in female sexual behavior.
11.6 A response through which an organism can avoid, escape from, or minimize an aversive stimulus; reduces the stressful effects of an aversive stimulus.	10.49 The region of the midbrain that surrounds the cerebral aqueduct; plays an essential role in various species-typical behaviors, including female sexual behavior.
11.7 The region of the prefrontal cortex at the base of the anterior frontal lobes.	10.50 The act of giving birth.
11.8 A motor disorder characterized by a relative lack of movement and lack of speech; caused by damage to the cingulate gyrus.	10.51 A long fiber bundle that connects portions of the amygdala with the hypothalamus.
11.9 A culturally determined rule that modifies the expression of emotion in a particular situation.	10.52 A nucleus in the ventral midbrain; plays an essential role in maternal behavior.
11.10 Difficulty in moving the facial muscles voluntarily; caused by damage to the face region of the primary motor cortex or its subcortical connections.	11.1 A group of subnuclei of the amygdala that receives sensory input, including information about the presence of odors and pheromones, and relays it to the medial basal forebrain and hypothalamus.
11.11 Lack of movement of facial muscles in response to emotions in people who have no difficulty moving these muscles voluntarily; caused by damage to the insular prefrontal cortex, subcortical white matter of the frontal lobe, or parts of the thalamus..	11.2 Nuclei of the amygdala that receive sensory information from the neocortex, thalamus, and hippocampus and send projections to the ventral striatum, dorsomedial nucleus of the thalamus, and the central nucleus.
11.12 A test often performed before brain surgery; verifies the functions of one hemisphere by testing patients while the other hemisphere is anesthetized.	11.3 The region of the amygdala that receives information from the basolateral division and sends projections to a wide variety of regions in the brain; involved in emotional responses.

11.13 James-Lange theory	12.5 detector
11.14 defensive behavior	12.6 correctional mechanism
11.15 threat behavior	12.7 negative feedback
11.16 submissive behavior	12.8 satiety mechanism
11.17 predation	12.9 intracellular fluid
12.1 homeostasis (*home ee oh **stay** sis*)	12.10 extracellular fluid
12.2 ingestive behavior (*in **jess** tiv*)	12.11 intravascular fluid
12.3 system variable	12.12 interstitial fluid
12.4 set point	12.13 isotonic

12.5 In a regulatory process, a mechanism that signals when the system variable deviates from its set point.	**11.13** A theory of emotion that suggests that behaviors and physiological responses are directly elicited by situations, and that feelings of emotions are produced by feedback from these behaviors and responses.
12.6 In a regulatory process, the mechanism that is capable of changing the value of the system variable.	**11.14** A species-typical behavior by which an animal defends itself against the threat of another animal.
12.7 A process whereby the effect produced by an action serves to diminish or terminate that action; a characteristic of regulatory systems.	**11.15** A stereotypical species-typical behavior that warns another animal that it may be attacked if it does not flee or show a submissive behavior.
12.8 A brain mechanism that causes cessation of hunger or thirst, produced by adequate and available supplies of nutrients or water.	**11.16** A stereotyped behavior shown by an animal in response to threat behavior by another animal; serves to prevent an attack.
12.9 The fluid contained within cells.	**11.17** Attack of one animal directed at an individual of another species, on which the attacking animal normally preys.
12.10 All body fluids outside cells: interstitial fluid, blood plasma, and cerebrospinal fluid.	**12.1** The process by which the body's substances and characteristics (such as temperature and glucose ievel) are maintained at their optimal level.
12.11 The fluid found within the blood vessels.	**12.2** Eating or drinking.
12.12 The fluid that bathes the cells, filling the space between the cells of the body (the "interstices").	**12.3** A variable that is controlled by a regulatory mechanism; for example, temperature in a heating system.
12.13 Equal in osmotic pressure to the contents of a cell. A cell placed in an isotonic solution neither gains nor loses water.	**12.4** The optimal value of the system variable in a regulatory mechanism.

12.14 hypertonic	12.23 diabetes insipidus (*in* **sipp** *i duss*)
12.15 hypotonic	12.24 osmometric thirst
12.16 hypovolemia (*hy poh voh* **lee** *mee a*)	12.25 osmoreceptor
12.17 nephron	12.26 OVLT (organum vasculosum of the lamina terminalis)
12.18 ureter (**your** *eh ter*)	12.27 volumetric thirst
12.19 aldosterone (*al* **dahs** *ter own*)	12.28 colloid (**kalh** *oyd*)
12.20 vasopressin (*vay zo* **press** *in*)	12.29 salt appetite
12.21 supraoptic nucleus (*sue pra* **op** *tik*)	12.30 renin (**ree** *nin*)
12.22 paraventricular nucleus	12.31 angiotensinogen (*ann gee oh ten* **sin** *oh jen*)

12.23 The loss of excessive amounts of water through the kidneys; caused by lack of secretion of vasopressin.	**12.14** The characteristic of a solution that contains enough solute that it will draw water out of a cell placed in it, through the process of osmosis.
12.24 Thirst produced by an increase in the osmotic pressure of the interstitial fluid relative to the intracellular fluid, thus producing cellular dehydration.	**12.15** The characteristic of a solution that contains so little solute that a cell placed in it will absorb water, through the process of osmosis.
12.25 A neuron that detects changes in the solute concentration of the interstitial fluid that surrounds it.	**12.16** Reduction in the volume of the intravascular fluid.
12.26 A circumventricular organ located anterior to the anteroventral portion of the third ventricle; served by fenestrated capillaries and thus lacks a blood-brain barrier.	**12.17** A functional unit of the kidney; extracts fluid from the blood and carries the fluid, through collecting ducts, to the ureter.
12.27 Thirst produced by hypovolemia.	**12.18** One of two tubes that carries urine from the kidneys to the bladder.
12.28 A soluble, gluelike substance made of large molecules that cannot penetrate cell membranes.	**12.19** A hormone of the adrenal cortex that causes the retention of sodium by the kidneys.
12.29 A craving for sodium chloride.	**12.20** A hormone secreted by the posterior pituitary gland that causes the kidneys to excrete a more concentrated urine, thus retaining water in the body.
12.30 A hormone secreted by the kidneys that causes the conversion of angiotensinogen in the blood into angiotensin.	**12.21** A hypothalamic nucleus that contains cell bodies of neurons that produce antidiuretic hormone and transport it through their axons to the posterior pituitary gland.
12.31 A protein in the blood that can be converted by renin to angiotensin.	**12.22** A hypothalamic nucleus that contains cell bodies of neurons that produce antidiuretic hormone and oxytocin and transport them through their axons to the posterior pituitary gland.

12.32 angiotensin (*ann gee oh **ten** sin*)	12.41 pylorus (*pie **lorr** us*)
12.33 saralasin (*sair a **lay** sin*)	12.42 duodenum (*doo oh **dee** num*)
12.34 losartan (*low **sar** tan*)	12.43 hepatic portal vein
12.35 nucleus of the solitary tract	12.44 atrial natriuretic peptide (*nay tree ur **ett** ik*)
12.36 subfornical organ (SFO)	13.1 glycogen (***gly** ko jen*)
12.37 median preoptic nucleus	13.2 insulin
12.38 zona incerta (*in **sir** ta*)	13.3 glucagon (***gloo** ka gahn*)
12.39 furosemide (*few **row** se myde*)	13.4 triglyceride (*try **gliss** er ide*)
12.40 esophageal fistula (*ee soff a **jee** ul **fiss** tew la*)	13.5 glycerol (***gliss** er all*)

12.41 The ring of smooth muscle at the junction of the stomach and duodenum that controls the release of the stomach contents.	**12.32** A peptide hormone that constricts blood vessels, causes the secretion of aldosterone, and produces thirst and a salt appetite.
12.42 The portion of the small intestine immediately adjacent to the stomach.	**12.33** A drug that blocks angiotensin receptors.
12.43 The vein that receives blood from the digestive system and passes it to the liver.	**12.34** A drug that blocks angiotensin receptors.
12.44 A peptide secreted by the atria of the heart when blood volume is higher than normal; increases water and sodium excretion, inhibits renin, vasopressin, and aldosterone secretion, and inhibits sodium appetite.	**12.35** A nucleus of the medulla that receives information from visceral organs and from the gustatory system.
13.1 A polysaccharide often referred to as *animal starch;* stored in liver and muscle; constitutes the short-term store of nutrients.	**12.36** A small organ located in the confluence of the lateral ventricles, attached to the underside of the fornix; contains neurons that detect the presence of angiotensin in the blood and excite neural circuits that initiate drinking.
13.2 A pancreatic hormone that facilitates entry of glucose and amino acids into the cell, conversion of glucose into glycogen, and transport of fats into adipose tissue.	**12.37** A small nucleus situated around the decussation of the anterior commissure; plays a role in thirst stimulated by angiotensin.
13.3 A pancreatic hormone that promotes the conversion of liver glycogen into glucose.	**12.38** An oblong extension of the midbrain reticular formation, extending from the midbrain to the medial diencephalon.
13.4 The form of fat storage in adipose cells; consists of a molecule of glycerol joined with three fatty acids.	**12.39** A diuretic; a drug that increases the production of urine.
13.5 A substance (also called glycerine) derived from the breakdown of triglycerides, along with fatty acids; can be converted by the liver into glucose.	**12.40** A diversion of the esophagus so that when an animal eats or drink, the substance does not reach the stomach.

13.6 fatty acid	13.15 capsaicin (*kap say sin*)
13.7 fasting phase	13.16 2,5-AM
13.8 absorptive phase	13.17 L-ethionine
13.9 sensory-specific satiety	13.18 sham feeding
13.10 conditioned flavor aversion	13.19 cholecystokinin (CCK) (*coal i sis toe ky nin*)
13.11 glucoprivation	13.20 ob mouse
13.12 lipoprivation	13.21 leptin
13.13 methyl palmoxirate (MP)	13.22 lateral parabrachial nucleus
13.14 mercaptoacetate (MA)	13.23 neuropeptide Y (NPY)

13.15 An ingredient in hot peppers that can destroy small, unmyelinated sensory axons that innervate the internal organs.	**13.6** A substance derived from the breakdown of triglycerides, along with glycerol; can be metabolized by most cells of the body except for the brain.
13.16 A drug that inhibits carbohydrate metabolism in the liver by making phosphate unavailable, thus blocking the production of ATP.	**13.7** The phase of metabolism during which nutrients are not available from the digestive system; glucose, amino acids, and fatty acids are derived from glycogen, protein, and adipose tissue during this phase.
13.17 A drug that inhibits carbohydrate metabolism in the liver by making adenosine unavailable, thus blocking the production of ATP.	**13.8** The phase of metabolism during which nutrients are absorbed from the digestive system; glucose and amino acids constitute the principal source of energy for cells during this phase, and excess nutrients are stored in adipose tissue in the form of triglycerides.
13.18 Feeding behavior of an animal with an open gastric or esophageal fistula that prevents food from remaining in the stomach.	**13.9** Satiety for a specific food that has been ingested recently in the absence of general satiety for all foods.
13.19 A hormone secreted by the duodenum that regulates gastric motility and causes the gallbladder (cholecyst) to contract; appears to provide a satiety signal transmitted to the brain through the vagus nerve.	**13.10** The avoidance of a relatively unfamiliar flavor that previously caused (or was followed by) illness.
13.20 A strain of mice whose obesity and low metabolic rate is caused by a mutation that prevents the production of leptin.	**13.11** A dramatic fall in the level of glucose available to cells; can be caused by a fall in the blood level of glucose or by drugs that inhibit glucose metabolism.
13.21 A hormone secreted by adipose tissue; decreased food intake and increased metabolic rate, primarily by inhibiting NPY-secreting neurons in the arcuate nucleus.	**13.12** A dramatic fall in the level of fatty acids available to cells; usually caused by drugs that inhibit fatty-acid metabolism.
13.22 A nucleus in the pons that receives gustatory information and information from the liver and digestive system and relays it to the forebrain.	**13.13** A drug that inhibits fatty-acid metabolism and produces lipoprivic hunger.
13.23 A peptide neurotransmitter whose release stimulates feeding, insulin and glucocorticoid secretion, decreased breakdown of triglycerides, and a decrease in body temperature.	**13.14** A drug that inhibits fatty-acid metabolism and produces lipoprivic hunger.

13.24
 	
    arcuate nucleus
	14.3
	
	    classical conditioning

13.25
 	
    galanin
    (*gal a nin*)
	14.4
	
	    Hebb rule

13.26
 	
    fenfluramine
    (FEN)
	14.5
	
	    instrumental conditioning

13.27
 	
    agouti mouse
	14.6
	
	    reinforcing stimulus

13.28
 	
    melanocortin-4 receptor
    (MC4-R)
	14.7
	
	    punishing stimulus

13.29
 	
    anorexia nervosa
	14.8
	
	    motor learning

13.30
 	
    bulimia nervosa
	14.9
	
	    long-term potentiation

14.1
 	
    perceptual learning
	14.10
	
	    hippocampal formation

14.2
 	
    stimulus-response learning
	14.11
	
	    entorhinal cortex

14.3 When a neutral stimulus is followed several times by an *unconditional stimulus* that produces a defensive or appetitive response (the *unconditional response*), the first stimulus (now called a *conditional stimulus*) itself evokes the response (now called a *conditional response*).	**13.24** A nucleus in the base of the hypothalamus that controls secretions of the anterior pituitary gland; contains NPY-secreting neurons involved in feeding and control of metabolism.
14.4 The hypothesis proposed by Donald Hebb that the cellular basis of learning involves strengthening of a synapse that is repeatedly active when the postsynaptic neuron fires.	**13.25** A peptide neurotransmitter whose release stimulates ingestion of fats.
14.5 A learning procedure whereby the effects of a particular behavior in a particular situation increase (reinforce) or decrease (punish) the probability of the behavior; also called *operant conditioning*.	**13.26** A drug that causes the release of serotonin and inhibits eating.
14.6 An appetitive stimulus that follows a particular behavior and thus makes the behavior become more frequent.	**13.27** A strain of mice whose yellow fur and obesity are caused by a mutation that causes the production of a peptide that blocks MC4 receptors in the brain.
14.7 An aversive stimulus that follows a particular behavior and thus makes the behavior become less frequent.	**13.28** A receptor normally stimulated by the hormone melanocortin; responsible for the production of melanin; also plays a role in control of appetite.
14.8 Learning to make a new response.	**13.29** A disorder that most frequently afflicts young women; exaggerated concern with overweight that leads to excessive dieting and often compulsive exercising; can lead to starvation.
14.9 A long-term increase in the excitability of a neuron to a particular synaptic input caused by repeated high-frequency activity of that input.	**13.30** Bouts of excessive hunger and eating, often followed by forced vomiting or purging with laxatives; sometimes seen in people with anorexia nervosa.
14.10 A forebrain structure of the temporal lobe, constituting an important part of the limbic system; includes the hippocampus proper (Ammon's horn), dentate gyrus, and subiculum.	**14.1** Learning to recognize a particular stimulus.
14.11 A region of the limbic cortex that provides the major source of input to the hippocampal formation.	**14.2** Learning to automatically make a particular response in the presence of a particular stimulus; includes classical and instrumental conditioning.

14.12 granule cell	14.21 AMPA receptor
14.13 dentate gyrus	14.22 dendritic spike
14.14 field CA3	14.23 protein kinase
14.15 pyramidal cell	14.24 CaM-KII
14.16 field CA1	14.25 tyrosine kinase
14.17 population EPSP	14.26 nitric oxide synthase
14.18 associative long-term potentiation	14.27 long-term depression
14.19 NMDA receptor	14.28 short-term memory
14.20 AP5	14.29 delayed matching-to-sample task

14.21 An ionotropic glutamate receptor that controls a sodium channel; when open, produces EPSPs.	14.12 A small, granular cell; those found in the dentate gyrus send axons to field CA3 of the hippocampus.
14.22 An action potential that occurs in the dendrite of some types of pyramidal cells.	14.13 Part of the hippocampal formation; receives inputs from the entorhinal cortex and projects to field CA3 of the hippocampus.
14.23 An enzyme that attaches a phosphate (PO_4) to a protein and thereby causes it to change its shape.	14.14 Part of the hippocampus; receives inputs from the dentate gyrus and projects to field CA1.
14.24 Type II calcium-calmodulin kinase, an enzyme that must be activated by calcium; may play a role in the establishment of long-term potentiation.	14.15 A category of large neurons with a pyramid shape; found in the cerebral cortex and Ammon's horn of the hippocampal formation.
14.25 A type of protein kinase that may play a role in the establishment of long-term potentiation.	14.16 Part of the hippocampus; receives inputs from field CA3 and projects out of the hippocampal formation via the subiculum.
14.26 An enzyme responsible for the production of nitric oxide.	14.17 An evoked potential that represents the EPSPs of a population of neurons.
14.27 A long-term decrease in the excitability of a neuron to a particular synaptic input caused by stimulation of the terminal button while the postsynaptic membrane is hyperpolarized or only slightly depolarized.	14.18 A long-term potentiation in which concurrent stimulation of weak and strong synapses to a given neuron strengthens the weak ones.
14.28 Memory for a stimulus that has just been perceived.	14.19 A specialized ionotropic glutamate receptor that controls a calcium channel that is normally blocked by Mg^{2+} ions; involved in long-term potentiation.
14.29 A task that requires the subject to indicate which of several stimuli has just been perceived.	14.20 2-amino-5-phosphonopentanoate; a drug that blocks NMDA receptors.

14.30 transcranial magnetic stimulation	14.39 conditioned reinforcer
14.31 nucleus basalis	14.40 conditioned punisher
14.32 MGm	15.1 anterograde amnesia
14.33 extinction	15.2 retrograde amnesia
14.34 self-stimulation	15.3 Korsakoff's syndrome
14.35 medial forebrain bundle (MFB)	15.4 confabulation
14.36 mesolimbic system	15.5 short-term memory
14.37 nucleus accumbens	15.6 long-term memory
14.38 mesocortical system	15.7 consolidation

14.39 A previously neutral stimulus that has been paired with an appetitive stimulus, which then itself becomes capable, of reinforcing a response.	14.30 Stimulation of the cortex by the magnetic field produced by alternating current passing through a coil placed against the skull; disrupts normal activity of the affected brain region.
14.40 A previously neutral stimulus that has been followed by an aversive stimulus, which then itself becomes capable of punishing a response.	14.31 A nucleus of the basal forebrain that contains most of the acetylcholine-secreting neurons that send axons to the neocortex; degenerates in patients with Alzheimer's disease.
15.1 Amnesia for events that occur after some disturbance to the brain, such as head injury or certain degenerative brain diseases.	14.32 The medial division of the medial geniculate nucleus; transmits auditory and somatosensory information to the lateral nucleus of the amygdala.
15.2 Amnesia for events that preceded some disturbance to the brain, such as a head injury or electroconvulsive shock.	14.33 With respect to classical conditioning, the reduction or elimination of a conditional response by repeatedly presenting the conditional stimulus without the unconditional stimulus.
15.3 Permanent anterograde amnesia caused by brain damage resulting from chronic alcoholism or malnutrition.	14.34 Making a response that causes the electrical stimulation of a particular region of the brain through an implanted electrode.
15.4 The reporting of memories of events that did not take place without the intention to deceive; seen in people with Korsakoff's syndrome.	14.35 A fiber bundle that runs in a rostral-caudal direction through the basal forebrain and lateral hypothalamus; electrical stimulation of these axons is reinforcing.
15.5 Immediate memory for events, which may or may not be consolidated into long-term memory.	14.36 A system of dopaminergic neurons whose cell bodies are located in the ventral tegmental area and whose terminal buttons are located in the nucleus accumbens, amygdala, lateral septum, hippocampus, and bed nucleus of the stria terminalis.
15.6 Relatively stable memory of events that occurred in the more distant past, as opposed to short-term memory.	14.37 A nucleus of the basal forebrain near the septum; receives dopamine-secreting terminal buttons from neurons of the ventral tegmental area and is thought to be involved in reinforcement and attention.
15.7 The process by which short-term memories are converted into long-term memories.	14.38 A system of dopaminergic neurons whose cell bodies are located in the ventral tegmental area and whose terminal buttons are located in the cerebral cortex and hippocampus.

15.8 priming	15.17 ventral angular bundle
15.9 declarative memory	15.18 recurrent collateral
15.10 nondeclarative memory	16.1 cerebrovascular accident
15.11 perirhinal cortex	16.2 aphasia
15.12 parahippocampal cortex	16.3 Broca's aphasia
15.13 working memory	16.4 function word
15.14 reference memory	16.5 content word
15.15 place cells	16.6 Broca's area
15.16 theta rhythm	16.7 agrammatism

15.17 The bundle of axons that conveys information from the hippocampal formation to the basolateral amygdala.	15.8 A phenomenon in which exposure to a particular stimulus automatically facilitates perception of that stimulus or related stimuli.
15.18 A branch of an axon leaving a particular region of the brain that turns back and forms synapses with neurons near the one that gives rise to it.	15.9 Memory that can be verbally expressed, such as memory for events in a person's past.
16.1 A "stroke"; brain damage caused by occlusion or rupture of a blood vessel in the brain.	15.10 Memory whose formation does not depend on the hippocampal formation; a collective term for perceptual, stimulus-response, and motor memory.
16.2 Difficulty in producing or comprehending speech not produced by deafness or a simple motor deficit; caused by brain damage.	15.11 A region of limbic cortex adjacent to the hippocampal formation that, along with the parahippocampal cortex, relays information between the entorhinal cortex and other regions of the brain.
16.3 A form of aphasia characterized by agrammatism, anomia, and extreme difficulty in speech articulation.	15.12 A region of limbic cortex adjacent to the hippocampal formation that, along with the perirhinal cortex, relays information between the entorhinal cortex and other regions of the brain.
16.4 A preposition, article, or other word that conveys little of the meaning of a sentence but is important in specifying its grammatical structure.	15.13 Memory of what has just been perceived and what is currently being thought about; consists of new information and related information that has recently been "retrieved" from long-term memory.
16.5 A noun, verb, adjective, or adverb that conveys meaning.	15.14 A form of long-term memory of stable conditions and contingencies in the environment; includes perceptual memory and stimulus-response memory.
16.6 A region of frontal cortex, located just rostral to the base of the left primary motor cortex, that is necessary for normal speech production.	15.15 A neuron of the hippocampus that becomes active when the animal is in a particular location in the environment.
16.7 One of the usual symptoms of Broca's aphasia; a difficulty in comprehending or properly employing grammatical devices, such as verb endings and word order.	15.16 EEG activity of 5-8 Hz; an important indication of the physiological state of the hippocampus.

16.8 anomia	16.17 circumlocution
16.9 apraxia of speech	16.18 prosody
16.10 Wernicke's area	16.19 pure alexia
16.11 Wernicke's aphasia	16.20 whole-word reading
16.12 pure word deafness	16.21 phonetic reading
16.13 transcortical sensory aphasia	16.22 surface dyslexia
16.14 autotopagnosia	16.23 phonological dyslexia
16.15 arcuate fasciculus	16.24 word-form dyslexia
16.16 conduction aphasia	16.25 spelling dyslexia

16.17 A strategy by which people with anomia find alternative ways to say something when they are unable to think of the most appropriate word.	**16.8** Difficulty in finding (remembering) the appropriate word to describe an object, action, or attribute; one of the symptoms of aphasia.
16.18 The use of changes in intonation and emphasis to convey meaning in speech besides that specified by the particular words; an important means of communication of emotion.	**16.9** Impairment in the ability to program movements of the tongue, lips, and throat required to produce the proper sequence of speech sounds.
16.19 Loss of the ability to read without loss of the ability to write; produced by brain damage.	**16.10** A region of auditory association cortex on the left temporal lobe of humans, which is important in the comprehension of words and the production of meaningful speech.
16.20 Reading by recognizing a word as a whole; "sight reading."	**16.11** A form of aphasia characterized by poor speech comprehension and fluent but meaningless speech.
16.21 Reading by decoding the phonetic significance of letter strings; "sound reading."	**16.12** The ability to hear, to speak, and (usually) to read and write without being able to comprehend the meaning of speech; caused by damage to Wernicke's area or disruption of auditory input to this region.
16.22 A reading disorder in which a person can read words phonetically but has difficulty reading irregularly spelled words by the whole-word method.	**16.13** A speech disorder in which a person has difficulty comprehending speech and producing meaningful spontaneous speech but can repeat speech; caused by damage to the region of the brain posterior to Wernicke's area.
16.23 A reading disorder in which a person can read familiar words but has difficulty reading unfamiliar words or pronounceable nonwords.	**16.14** Inability to name body parts or to identify body parts that another person names.
16.24 A disorder in which a person can read a word only after spelling out the individual letters.	**16.15** A bundle of axons that connects Wernicke's area with Broca's area; damage causes conduction aphasia.
16.25 An alternative name for word-form dyslexia.	**16.16** An aphasia characterized by inability to repeat words that are heard but normal speech and the ability to comprehend the speech of others.

16.26 direct dyslexia	17.5 hallucination
16.27 phonological dysgraphia	17.6 negative symptom
16.28 orthographic dysgraphia	17.7 chlorpromazine
16.29 developmental dyslexia	17.8 clozapine
16.30 planum temporale	17.9 tardive dyskinesia
17.1 schizophrenia	17.10 supersensitivity
17.2 positive symptom	17.11 epidemiology
17.3 thought disorder	17.12 seasonality effect
17.4 delusion	17.13 latitude effect

17.5 Perception of a nonexistent object or event.	16.26 A language disorder caused by brain damage in which the person can read words aloud without understanding them.
17.6 A symptom of schizophrenia characterized by the absence of behaviors that are normally present: social withdrawal, lack of affect, and reduced motivation.	16.27 A writing disorder in which the person cannot sound out words and write them phonetically.
17.7 A dopamine receptor blocker; a most commonly prescribed antischizophrenic drug.	16.28 A writing disorder in which the person can spell regularly spelled words but not irregularly spelled ones.
17.8 An "atypical" antipsychotic drug; blocks D_4 receptors in the nucleus accumbens.	16.29 A reading difficulty in a person of normal intelligence and perceptual ability; of genetic origin or caused by prenatal or perinatal factors.
17.9 A movement disorder that can after prolonged treatment with antipsychotic medication, characterized by involuntary movements of the face and neck.	16.30 A region of the superior temporal lobe; normally larger in the left hemisphere.
17.10 The increased sensitivity of neurotransmitter receptors; caused by damage to the afferent axons or long-term blockage of neurotransmitter release.	17.1 A serious mental disorder characterized by disordered thoughts, delusions, hallucinations, and often bizarre behaviors.
17.11 The study of the distribution and causes of diseases in populations.	17.2 A symptom of schizophrenia evident by its presence: delusions, hallucinations, or thought disorders.
17.12 The increased incidence of schizophrenia in people born during late winter and early spring.	17.3 Disorganized, irrational thinking.
17.13 The increased incidence of schizophrenia in people born far from the equator.	17.4 A belief that is clearly in contradiction to reality.

17.14 major affective disorder	17.23 monoamine hypothesis
17.15 bipolar disorder	17.24 5-HIAA
17.16 unipolar depression	17.25 tryptophan depletion procedure
17.17 tricyclic antidepressant	17.26 subsensitivity
17.18 specific serotonin reuptake inhibitor (SSRI)	17.27 seasonal affective disorder
17.19 electroconvulsive therapy (ECT)	17.28 summer depression
17.20 lithium	17.29 phototherapy
17.21 phosphoinositide system	18.1 anxiety disorder
17.22 carbamazepine	18.2 panic disorder

17.23 A hypothesis that states that depression is caused by a low level of activity of one or more monoaminergic synapses.	17.14 A serious mood disorder; includes unipolar depression and bipolar disorder.
17.24 A breakdown product of the neurotransmitter serotonin (5-HT).	17.15 A serious mood disorder characterized by cyclical periods of mania and depression.
17.25 A procedure involving a low tryptophan diet and a tryptophan-free amino acid "cocktail" that lowers brain tryptophan and consequently decreases the synthesis of 5-HT.	17.16 A serious mood disorder that consists of unremitting depression or periods of depression that do not alternate with periods of mania.
17.26 Decreased sensitivity of neurotransmitter receptors; a compensatory response to their prolonged stimulation.	17.17 A class of drugs used to treat depression; inhibits the reuptake of norepinephrine and serotonin; named for the molecular structure.
17.27 A mood disorder characterized by depression, lethargy, sleep disturbances, and craving for carbohydrates during the winter season when days are short.	17.18 A drug that inhibits the reuptake of serotonin without affecting the reuptake of other neurotransmitters.
17.28 A mood disorder characterized by depression, sleep disturbances, and loss of appetite.	17.19 A brief electrical shock, applied to the head, that results in an electrical seizure; used therapeutically to alleviate severe depression.
17.29 Treatment of seasonal affective disorder by daily exposure to bright light.	17.20 An element; lithium carbonate is used to treat bipolar disorder.
18.1 A psychological disorder characterized by tension, overactivity of the autonomic nervous system, expectation of an impending disaster, and continuous vigilance for danger.	17.21 A biochemical pathway responsible for the production of several second messengers.
18.2 A disorder characterized by episodic periods of symptoms such as shortness of breath, irregularities in heartbeat, and other autonomic symptoms, accompanied by intense fear.	17.22 A drug (trade name: Tegretol) used to treat seizures originating from a focus, generally in the medial temporal lobe.

18.3 anticipatory anxiety	18.12 fragile X syndrome
18.4 agoraphobia	18.13 stress
18.5 obsessive-compulsive disorder	18.14 stressor
18.6 obsession	18.15 stress response
18.7 compulsion	18.16 fight-or-flight response
18.8 Tourette's syndrome	18.17 glucocorticoid
18.9 Sydenham's chorea	18.18 corticotropin-releasing factor (CRF)
18.10 autistic disorder	18.19 adrenocorticotropic hormone (ACTH)
18.11 phenylketonuria	18.20 posttraumatic stress disorder

18.12 A genetic disorder caused by a faulty gene on the X chromosome; the leading genetic cause of mental retardation.	**18.3** A fear of having a panic attack; may lead to the development of agoraphobia.
18.13 A general, imprecise term that can refer either to a stress response or to a stressor (stressful situation).	**18.4** A fear of being away from home or other protected places.
18.14 A stimulus (or situation) that produces a stress response.	**18.5** A mental disorder characterized by obsessions and compulsions.
18.15 A physiological reaction caused by the perception of aversive or threatening situations.	**18.6** An unwanted thought or idea with which a person is preoccupied.
18.16 A species-typical response preparatory to fighting or fleeing; thought to be responsible for some of the deleterious effects of stressful situations on health.	**18.7** The feeling that one is obliged to perform a behavior, even if one prefers not to do so.
18.17 One of a group of hormones of the adrenal cortex that are important in protein and carbohydrate metabolism, secreted especially in times of stress.	**18.8** A neurological disorder characterized by tics and involuntary vocalizations and sometimes by compulsive uttering of obscenities and repetition of the utterances of others.
18.18 A hypothalamic hormone that stimulates the anterior pituitary gland to secrete ACTH (adrenocorticotrophic hormone).	**18.9** An autoimmune disease that attacks parts of the brain including the basal ganglia and produces involuntary movements and often the symptoms of obsessive-compulsive disorder.
18.19 A hormone released by the anterior pituitary gland in response to CRF; stimulates the adrenal cortex to produce glucocorticoids.	**18.10** A chronic disorder whose symptoms include failure to develop normal social relations with other people, impaired development of communicative ability, lack of imaginative ability, and repetitive, stereotyped movements.
18.20 A psychological disorder caused by exposure to a situation of extreme danger and stress; symptoms include recurrent dreams or recollections; can interfere with social activities and a feeling of hopelessness.	**18.11** A hereditary disorder caused by the absence of an enzyme that converts the amino acid phenylalanine to tyrosine; causes brain damage unless a special diet is implemented soon after birth.

18.21 psychoneuroimmunology	19.2 withdrawal symptoms
18.22 natural killer cell	19.3 substance dependence
18.23 antigen	19.4 substance abuse
18.24 antibody	19.5 negative reinforcement
18.25 B-lymphocyte	19.6 dynorphin
18.26 immunoglobulin	19.7 conditioned place preference
18.27 T-lymphocyte	19.8 naloxone
18.28 cytokine	19.9 pimozide
19.1 tolerance	19.10 antagonist-precipitated withdrawal

19.2 The appearance of symptoms opposite to those produced by a drug when the drug is suddenly no longer taken; caused by the presence of compensatory mechanisms.	18.21 The branch of neuroscience involved with interactions between environmental stimuli, the nervous system, and the immune system.
19.3 A maladaptive pattern of substance abuse that includes taking increasing doses of the drug or other signs of addiction.	18.22 A white blood cell that destroys cancer cells and cells infected by viruses.
19.4 A maladaptive pattern of substance use short of addiction that interferes with a person's health or social situation.	18.23 A protein present on a microorganism that permits the immune system to recognize it as an invader.
19.5 The removal or reduction of an aversive stimulus that is contingent on a particular response, with an attendant increase in the frequency of that response.	18.24 A protein produced by a cell of the immune system that recognizes antigens present on invading microorganisms.
19.6 An endogenous opioid; the natural ligand for kappa opiate receptors.	18.25 A white blood cell that originates in the bone marrow; part of the immune system.
19.7 The learned preference for a location in which an organism encountered a reinforcing stimulus, such as food or a reinforcing drug.	18.26 An antibody released by B-lymphocytes that bind with antigens and help destroy invading microorganisms.
19.8 A drug that blocks mu opiate receptors; antagonizes the reinforcing and sedative effects of opiates.	18.27 A white blood cell that originates in the thymus gland; part of the immune system.
19.8 A drug that blocks dopamine receptors.	18.28 A category of chemicals released by certain white blood cells when they detect the presence of an invading microorganism ; causes other white blood cells to proliferate and mount an attack against the invader.
19.10 Sudden withdrawal from long-term administration of a drug caused by cessation of the drug and administration of an antagonistic drug.	19.1 The fact that increasingly large doses of drugs must be taken to achieve a particular effect; caused by compensatory mechanisms that oppose the effect of the drug.

19.11 CREB	
19.12 drug discrimination procedure	

	19.11 Cyclic AMP-responsive element-binding protein; a nuclear protein to which cyclic AMP can bind and affect the activity of a gene or set of genes.
	19.12 An experimental procedure in which an animal shows, through instrumental conditioning, whether the perceived effects of two drugs are similar.